MODELING AND NONLINEAR ROBUST CONTROL OF DELTA-LIKE PARALLEL KINEMATIC MANIPULATORS

MODELING AND NONLINEAR ROBUST CONTROL OF DELTA-LIKE PARALLEL KINEMATIC MANIPULATORS

JONATAN MARTIN ESCORCIA HERNÁNDEZ
Division of Engineering, Polytechnic University of Tulancingo
Tulancingo, Mexico

AHMED CHEMORI
LIRMM, University of Montpellier, CNRS
Montpellier, France

HIPÓLITO AGUILAR SIERRA
Faculty of Engineering, La Salle University Mexico
Ciudad de Mexico, Mexico

ELSEVIER

ACADEMIC PRESS
An imprint of Elsevier

Academic Press is an imprint of Elsevier
125 London Wall, London EC2Y 5AS, United Kingdom
525 B Street, Suite 1650, San Diego, CA 92101, United States
50 Hampshire Street, 5th Floor, Cambridge, MA 02139, United States
The Boulevard, Langford Lane, Kidlington, Oxford OX5 1GB, United Kingdom

ISBN: 978-0-323-96101-1

For information on all Academic Press publications
visit our website at https://www.elsevier.com/books-and-journals

Publisher: Mara E. Conner
Acquisitions Editor: Sonnini R. Yura
Editorial Project Manager: Lira Faurillo
Production Project Manager: Maria Bernard
Cover Designer: Matthew Limbert

Typeset by VTeX

Working together
to grow libraries in
developing countries

www.elsevier.com • www.bookaid.org

Contents

4. Proposed robust control solutions

5. Simulation and real-time experimental results

About the authors

Jonatan Martin Escorcia Hernández received his B.Sc. in Robotic Engineering, M.Sc. in Automation and Control, and Ph.D. in Optomechatronics from the Polytechnic University of Tulancingo (UPT), Tulancingo de Bravo, Mexico in 2013, 2017, and 2020, respectively. His research interests include modeling, mechanical design, and nonlinear control of robotics systems.

Ahmed Chemori received his M.Sc. and Ph.D. degrees both in automatic control from the Grenoble Institute of Technology, Grenoble, France, in 2001 and 2005, respectively. He has been a Postdoctoral Fellow with the Automatic Control Laboratory, Grenoble, France, in 2006. He is currently a tenured Research Scientist in automatic control and robotics with the Montpellier Laboratory of Informatics, Robotics and Microelectronics (LIRMM). His research interests include nonlinear, adaptive, robust and predictive control and their real-time applications in complex robotic systems.

Hipólito Aguilar Sierra received the B.Sc. degree in Mechatronics Engineering from UPIITA-IPN in 2009; and M.Sc. and Ph.D. degrees both in Automatic Control from the CINVESTAV Zacatenco, Mexico City, Mexico, in 2011 and 2016, respectively. He is currently a Full-time professor at Faculty of Engineering from the La Salle Mexico University. His research interests include Medical robots, Rehabilitation robots, Exoskeleton robotics and Nonlinear control.

Preface

In recent decades, the potentially higher accuracy, stiffness, and dynamics of parallel kinematic manipulators (PKM), also called parallel kinematic machines or parallel robots for short, have been catching the increasing attention of many researchers around the world, compared to serial manipulators. These special characteristics are due to the closed-loop kinematic structure that allows them to deal with multiple applications such as Pick-and-Place, surgical assistance, haptic devices, 3D printing, motion simulators, machining tasks, etc. However, many closed kinematic chains and configurations make PKMs highly nonlinear and complex systems. Besides, these manipulators usually have parameter uncertainties and non-parametric uncertainties. Moreover, in some cases, there are mechanical configurations where the number of actuators is higher than the number of degrees of freedom (DOF) resulting in what we call actuation redundancy. This redundant actuated (RA) configuration can cause the generation of internal forces, and due to the complexity of the system, it is not possible to predict the magnitude of these forces. Accordingly, it is necessary to restrict the dynamics of the robot in order to avoid damaging the mechanical structure. To successfully achieve one of these applications, even in the presence of the aforementioned issues, it is necessary to design and implement an advanced control system that can guarantee good tracking performances. This book presents the modeling and control algorithms for two PKMs with delta-like architecture. The first one is the widely known 3-DOF Delta robot and a novel 5-DOF RA-PKM called SPIDER4, which is designed to perform machining tasks such as milling or drilling. The kinematic and dynamic models for the Delta robot and SPIDER4 are described in detail, the inverse kinematics model is obtained by a closed-loop formula, the forward kinematics model is calculated by the virtual sphere's intersection algorithm, and finally, the dynamic model is obtained based on the virtual work principle using a delta-like structure as a simplified manipulator. In this work we propose two control solutions based on the Robust Integral of the Sign of Error (RISE) to control the motion of the robot. In the first one, we incorporate a feedforward compensation term into the control loop based on the inverse dynamic model of the manipulators; in the second one, feedback adaptive gains are added into the control loop to compensate for the perturbations produced by changes in the payload as may occur pick-and-place tasks and the contact forces resulting from the machining tasks. Finally, the proposed control schemes as well as the mathematical models are verified through numerical simulations and real-time experiments.

Jonatan Martin Escorcia Hernández, Ahmed Chemori, and
Hipólito Aguilar Sierra

CHAPTER

1

Introduction

1.1 Classification of robotic manipulators

Robots are mechatronic systems formed by mechanical, electrical, and electronic elements controlled by commands generated by a control system (computer, PLC, microcontroller, etc.). Nowadays, they are present in oceans, the sky, schools, factories, hospitals, and homes doing productive or leisure activities. According to The Robotics Institute of America, *a robot is a re-programmable multi-functional manipulator designed to move materials, parts tools, or specialized devices through variable programmed motions for the performance of a variety of tasks* [126]. Manipulation robotics is a branch of robotics aiming at carrying out tasks developed by human arms to handle and position objects. Currently, robotic manipulators play an important role in industrial automation, preserving both quantity and quality in production lines. They can be reconfigured to perform different automated tasks, such as spray painting, welding, material handling, or components assembly.

In several works reported in the literature, robotic manipulators are classified according to their kinematic configuration. There are three kinematic configurations for robotic manipulators: serial, parallel, and hybrid. Each kinematic architecture has certain advantages over the others in their mechanical construction that makes it attractive for carrying out specific tasks; let us expound each one of them.

1.1.1 Serial manipulators

Serial robots are the most extended manipulators in industrial applications. They have an open kinematic chain formed by several links connected to each other by prismatic or revolute joints. Usually, serial manipulators have actuators in all joints. The end-effector in a serial manipulator is located at the end of the kinematic chain. One example of a serial manip-

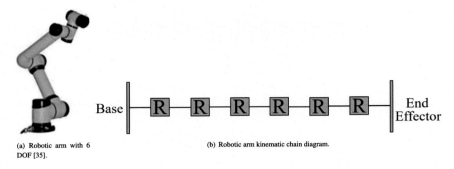

(a) Robotic arm with 6
DOF [35].

(b) Robotic arm kinematic chain diagram.

FIGURE 1.1   Exemplification of a serial manipulator with its kinematic configuration.

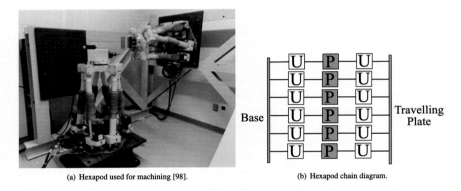

(a) Hexapod used for machining [98].

(b) Hexapod chain diagram.

FIGURE 1.2   Exemplification of a PKM with its kinematic configuration.

ulator is the robotic arm imitating the human arm parts (shoulder, elbow, and wrist). Fig. 1.1(a) shows a six Degrees of Freedom (DOF) serial robotic arm; its kinematic configuration is illustrated in Fig. 1.1(b), where the gray boxes with a letter R denote the active revolute joints.

1.1.2 Parallel kinematic manipulators

Parallel Kinematic Manipulators (PKMs), also known as Parallel Robots or Parallel Kinematic Machines, are robotic devices designed as a solution for the limitations presented in Serial Manipulators in terms of stiffness and precision at high accelerations. According to [80], *a generalized Parallel Manipulator is a closed kinematic loop mechanism, whose end-effector is connected to the base through at least two independent kinematic chains.* Fig. 1.2(a) shows a 6-DOF Hexapod being a typical example of a PKM. The manipulator's kinematic configuration is illustrated in Fig. 1.2(b), where the gray boxes with a letter P represent active prismatic joints. In contrast, the white boxes with a letter U symbolize passive universal joints.

1.1.3 Serial versus parallel manipulators

Serial manipulators and PKMs have various advantages and disadvantages due to their kinematic construction, which are listed below [70].

1.1.3.1 Advantages of serial manipulators

The main appreciated features of serial robots are:

- Large workspace volume.
- Good Dexterity proprieties.

1.1.3.2 Drawbacks of serial manipulators

In contrast to the above features, the main drawbacks of serial manipulators are as follows [70], [57].

- Low stiffness caused by their open-loop kinematic structure.
- High inertia due to the masses distributed over the entire kinematic chain (actuators, transmission components, etc.).
- Accumulative positioning errors due to the layout of the series segments.
- Increased fatigue and wear of the active joints.

1.1.3.3 Advantages of parallel kinematic manipulators

The closed-loop kinematics in PKMs entails the following advantages over their serial counterparts [80], [70], [57]:

- High mechanical stiffness since the payload is handled by all the kinematic chains.
- High load capacity.
- Low moving mass and inertia.
- Ability to perform very high dynamic movements (high speeds and accelerations).
- Possibility of positioning the actuators directly on the fixed base or very close to it; this characteristic has the following positive consequences:
  - Wide choice of motors and gear-heads since their mass does not influence on the mass and inertia of the moving parts.
  - Significant simplification of connection problems between motors, sensors and controller (more straightforward and more reliable wiring).
  - Comfortable cooling of the actuators, and therefore reduction of precision problems due to expansion and high potential power.

TABLE 1.1   Proprieties comparison between serial and parallel manipulators.

Proprieties	Serial Manipulators	Parallel Manipulators
Stiffens	Low	High
Workspace	Large	Small
Actuator's location	In the joints	In the base
Inertia forces and stiffness	High and less respectively	Less and high respectively
Payload/weight ratio	Low	High
Inverse kinematics	Difficult and complex	Straightforward and unique
Forward kinematics	Straightforward and unique	Difficult and complex
Singularities	Inverse kinematic	Inverse, forward, and combined

1.1.3.4 *Drawbacks of parallel kinematic manipulators*

Compared to serial robots, the disadvantages of parallel robots can be summarized as follows:

- Limited working volume workspace compared to the total volume of the mechanism.
- Forward Kinematic Models (FKM) are sometimes very challenging to determine.
- Presence of singularities which may lead to a loss of control of the mobile platform, or even to a deterioration of the mechanics. This is the most critical point when designing a machine with a parallel architecture.
- The use of many passive links may induce wobbles, making the robot's behavior difficult to model. However, they must be mastered to define dangerous zones near the singularities and to improve the precision of the robot.
- Internal forces generation that may produce mechanical damages in the case of Redundantly Actuated (RA) PKMs.

In accordance with [12], [107], and the arguments mentioned above, one can summarize the proprieties of Serial and Parallel Manipulators as illustrated in Table 1.1.

1.1.4 Hybrid manipulators

A Hybrid Manipulator can be defined as a combination of two mechanisms where one of them is based on a serial structure, and the other one is parallel. This kind of architecture can be divided into two branches:

- *Serial positioning device with a parallel wrist:* This configuration associates serial with parallel structures in the following form: the serial mechanism is responsible for positioning, while the parallel mechanism is used for the end-element orientation. In this category of hybrid manipu-

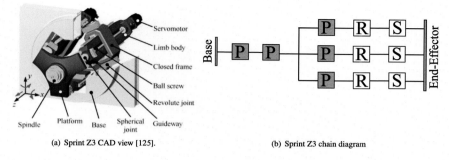

(a) Sprint Z3 CAD view [125].

(b) Sprint Z3 chain diagram

FIGURE 1.3 Hybrid manipulator with serial carrier and parallel wrist.

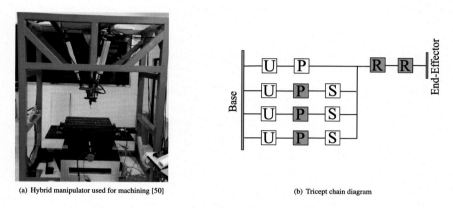

(a) Hybrid manipulator used for machining [50]

(b) Tricept chain diagram

FIGURE 1.4 Tricept hybrid manipulator with a parallel carrier and a serial wrist.

lators, we can mention the Sprint Z3 from DS Technology developed for aerospace applications [33]. The carrier mechanism uses two prismatic actuators to move the wrist in the plane x-y. The last one provides two rotational motions and one translation along z-axis; this manipulator is illustrated in Fig. 1.3, where the white boxes with a letter S represent passive spherical joints.

- *Parallel positioning device with serial wrist:* In this case, the carrier mechanism is the Parallel kinematic structure coupled with a Serial wrist responsible for orienting the end-effector. An example of such configuration is the 3T-2R DOF manipulator depicted in Fig. 1.4; in which three Universal-Prismatic-Spherical kinematic chains perform the three translational movements, while a serial wrist mechanism performs the two rotational movements. The manipulator utilizes a passive kinematic chain (the U-P chain) to constrain the parallel structure's platform [50].

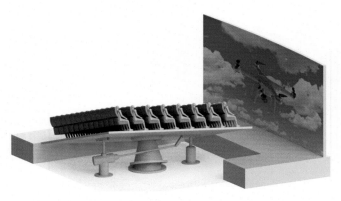

FIGURE 1.5   Spatial platform patented by James E. Gwinnett, illustration adapted from [59].

1.2 Overview of Parallel Kinematic Manipulators (PKMs)

Having explained the concepts related to robotic manipulators in a general form, let us now focus about PKMs, starting with their historical development and continuing with their potential applications.

1.2.1 Historical review of PKMs

According to [18], in the twentieth century, James E. Gwinnett designed a moving platform for the entertainment industry (a dynamic cinema). The patent for this system was requested in 1928. Unfortunately, the designed mechanism was never built [59] (see Fig. 1.5). After in 1938, Willard L.V. Pollard designed the first industrial parallel robot, intended for spray painting operations. The design consists of a system of five degrees of freedom, composed of three link sets. Each set is formed by a proximal and a distal arms joined through universal joints. Three actuators mounted on the base command the tool's position, while its orientation is controlled by the two other actuators located on the base, transmitting the movement to the tool using flexible cables. The parallel robot of Willard L.V. Pollard is shown in Fig. 1.6 [100]. In 1947, Dr. Eric Gough designed the first octahedral hexapod type platform with variable length sides. Dr. Gough called it the universal testing machine or universal platform. This system was invented to resolve the problems of air-landing loads; that is, the Gough platform tried to simulate an airplane's landing process. This machine was used to check the tires of Dunlop house under loads applied along different axes. The Gough platform has been one of the mechanisms that has achieved the greatest recognition in parallel robotics. Fig. 1.7 presents an example of the Gough platform [58]. In 1965 Stewart published an article in which he designed a 6-DOF platform intended to be used as a flight

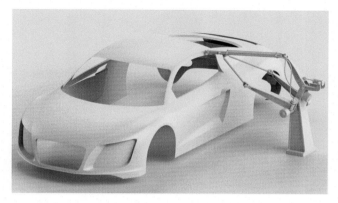

FIGURE 1.6 Illustration of Willard L.V. Pollard parallel mechanism illustration adapted from [100].

FIGURE 1.7 View of Gough platform [72].

simulator. His design was different from the one proposed by Gough. Ironically, the Gough platform is often referred to as the Stewart platform. Fig. 1.8 depicts a sketch of the Steward platform [119]. However, Stewart is not considered the creator of the first flight simulator based-on PKM. Indeed, in the '60s, the US engineer Klaus Cappel designed and built the first functional flight simulator. His design was based on an octahedral hexapod that with the same kinematic arrangement proposed years ago by Dr. Gough. The Cappel patent was published in 1964; however, he was not aware of the invention of Dr. Gough and Stewart's paper,

FIGURE 1.8   Illustration of the Stewart platform, illustration adapted from [119].

FIGURE 1.9   View of a flight simulator based on the prototype of Klaus Cappel [76].

which was not yet published. Fig. 1.9 shows the first flight simulator built by Klaus Cappel [27]. Since that about twenty years passed without any significant progress in the subject of PKMs until, in the mid-80s, when Reymond Clavel, professor of the École Polytechnique Fédélale de Lausanne (EPFL), introduces the 3T-1R-DOF Delta PKM designed for Pick-and-Place (P&P) tasks [37]. The Delta robot is mainly composed of three

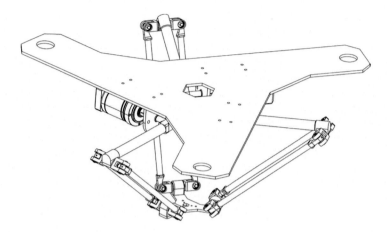

FIGURE 1.10   An illustration of the Clavel's Delta robot [5].

Revolute-Spherical-Spherical (RSS) kinematic chains using mechanisms based on parallelograms restraining entirely the orientation of the traveling plate, resulting in only translational movements, and an interdependent Rotational-Universal-Prismatic-Universal (RUPU) chain performing the rotational movement of the Delta robot. This PKM can reach high accelerations (up to 50G) thanks to its lightweight mechanism. The Delta robot is one of the most successful PKMs in history ever with many modifications around the world. In 1999, Pr. Clavel was awarded the "Golden Robot Award" (sponsored by ABB) for his innovative creation. Fig. 1.10 shows the original design of the Delta PKM [121].

1.2.1.1 *Other variants of Delta PKM over time*

Several modifications have been made to the original prototype of Pr. Clavel. In [94], F. Pierrot extended the concept of the Delta PKM from 3-DOF to 6-DOF, leading to the creation of "Hexa" PKM. Unlike the Delta robot, the Hexa robot can orient its traveling plate thanks to the fact that it has six kinematic chains where an individual motor actuates each one. In order to eliminate the RUPU chain present in the Delta robot, some mechanical solutions based on articulated traveling plates were proposed. The first prototype, called "H4," consists of four kinematic chains connected to an articulated traveling plate composed of three pieces [42]. Later, in [70], the traveling plate of H4 was customized, adding a gear mechanism to improve its performance. Notwithstanding, this PKM has some drawbacks such as the abundance of singularities and the possible internal collisions, so a careful selection of the positions was necessary to avoid them. To overcome the issues related to H4, a new family of PKMs called I4 was proposed in [69]. The main difference between I4 and H4

PKMs is the replacement of pivot joints with prismatic ones and gears with rack-and-pinion mechanisms. This family of PKMs has two variants I4L and I4R, which were actuated by linear and rotational actuators. However, the drawback of an articulated traveling plate using prismatic joints is its short life if the PKM is frequently operated at high-speeds. To overcome this limitation, in [83], a new PKM called Par4 was developed with a new concept of articulated traveling plate consisting of four main parts linked by two rotational joints. Two mechanisms, one using a gear assembly and another with pulleys and belts, were proposed to amplify the produced rotational movement. Nowadays, Par4 is commercialized by Adept Technologies under Quattro's name, considered the fastest industrial PKM in the world [12]. A modified version of Par4 called Heli4 was presented in [84]. The difference between this PKM and Par4 is the traveling plate composed of two parts connected through a helical screw that produces vertical axis rotation. Penta Robotics commercializes Heli4 under the name of Veloce [113]. In [7], a 2-DOF PKMs named Par2 was proposed for very high-speed motion on the x, z plane. This PKM is formed by two sets of active and passive kinematic chains; these last ones have the function to restrict the movement in only one plane. It has been reported that this PKM can reach high accelerations up to 50G [97]. In [56], a novel 2-DOF PKM concept called IRSBOT-2 was presented, constructed of only two active kinematic chains. Therefore, it is less subject to uncontrolled parasitic effects that may be produced by the passive kinematic chains of Par2. In the literature, there have been reported other prominent PKMs which make use of a Redundantly Actuated (RA) configuration. They are R4, with 3-DOF and four actuators, destined for applications that require extreme accelerations (up to 100G) [43], [86] SPIDER4, being the first delta-like PKM dedicated to machining operations, and T3KR dedicated to P&P applications. This last PKM shares several similarities with Par4; however, the rotational movement is performed by an independent actuator located at the traveling plate. Fig. 1.11 presents a timeline with all the examples presented in this section.

1.2.2 Main applications of parallel kinematic manipulators

Thanks to their main features as improved stiffness, high dynamics, and potentially improved accuracy, PKMs have gained a great interest in industries where such features play a crucial role. Therefore, let us explain some of their potential applications.

1.2.2.1 Pick-and-Place (P&P) tasks

The concept of P&P tasks is basically to pick an object from an initial position and place it to another position. This action is essential in assembly lines or in packaging production lines. Lightweight PKMs as delta-like manipulators are the best candidates to develop high-speed P&P opera-

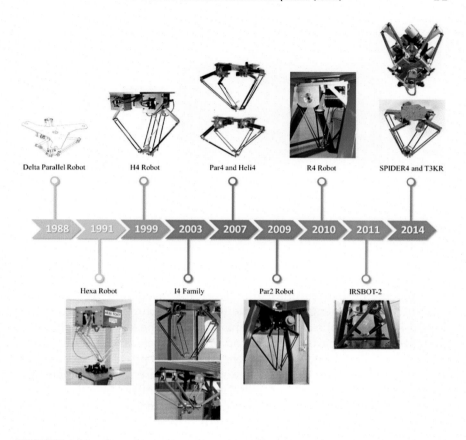

FIGURE 1.11   Timeline of some PKMs based on the Delta robot.

tions thanks to their improved stiffness and their high dynamic response. The original Delta PKM developed by Pr. Clavel was designed in the first instance to manipulate light mass objects in packaging lines. As shown before, the original Delta robot has been modified to satisfy the greatest P&P requirements, such as high positioning precision under extremely high-speed demand, improved orientation capabilities, higher payload capacity, and an extended operational workspace. Fig. 1.12 shows a food line assembly with Delta PKMs from the Swiss company Demaurex, and Adept Quattro being the fastest commercialized PKM.

1.2.2.2 *Machining operations*

Machining is the broad term used to describe the removal of material from a workpiece and is one of the most critical manufacturing processes. Machining operations can be applied to metallic and non-metallic mate-

(a) Delta PKM in Demaurex food industry [29]   (b) Adept Quattro [22]

FIGURE 1.12   Two examples of PKMs for P&P tasks.

rials such as polymers, wood, ceramics, composites, and other materials. The most common machining operations are milling, turning, and drilling. These operations require high precision in the positioning of the cutting tool and the desired cutting path. Hence the machine tools should satisfy these requirements. PKMs, due to their structure conformed by the closed-loop kinematic chain mechanism, have some key advantages over their serial counterparts in terms of accuracy, stiffness, and moving masses and inertias. The closed kinematic chains of PKMs provide better support to the cutting spindle due to the existence of more support points, and therefore the robot is less susceptible to errors produced by contact forces [136]. However, the main drawbacks of PKMs lie in their small workspace and extensive singular configurations [80], [91]. Some notable examples of PKMs for machining are: the Variax Hexacenter developed by Giddings & Levis company with 6-DOFs based on the Gough platform [80], Orthoglide PKM, having 3 translational DOFs with a fixed orientation [30], ARROW PKM which has 5-DOFs (3T-2R) [113], HexaM of Toyoda with 6-DOFs (3T-3R) [96], P800/P200 from Metrom, Okuma Cosmo Center and The Octahedral Hexapod from Ingersoll with 5-DOFs [80]. Fig. 1.13 illustrates two of these examples.

1.2.2.3 Coordinate measuring machines

Measuring is another potential application that PKMs can execute successfully due to the good accuracy provided by their closed-loop kinematic construction, increasing also the structural stability of the machine. Traditionally, serial devices have been used to perform this operation. However, errors caused by geometric offsets and deformations by load effects, and a serial structure's low stiffness may affect the precision of the measurements [139]. An example of a coordinate measuring machine is illustrated in Fig. 1.14.

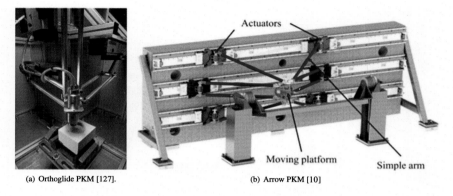

(a) Orthoglide PKM [127].   (b) Arrow PKM [10]

FIGURE 1.13   Two examples of PKMs for machining operations.

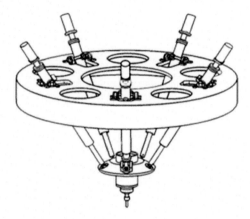

FIGURE 1.14   PKM used for coordinate measuring [34].

1.2.2.4 Medical applications

Thanks to their closed-loop architecture, some PKMs as Gough-Stewart platform or delta-like manipulators have been successfully integrated in this area. PKMs present desired features that make them attractive solutions in the medical field [130]. For instance, robotic-assisted surgery requires devices that take up as little space as possible and guaranteeing precise positioning of the surgical instruments [117]. We can mention some examples, for instance, the surgiscope is a tool-holder device developed by the Intelligent Surgical Instruments & Systems company (ISIS). It is based on a delta-like manipulator dedicated to microscope applications in neurosurgery and contains a microscope on the traveling plate whose mass is approximately 70 kg [20]. Moreover, the applications of the Delta PKM in medicine are not only limited to the positioning of tools; for example, in

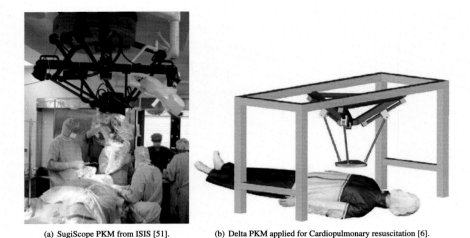

(a) SugiScope PKM from ISIS [51].   (b) Delta PKM applied for Cardiopulmonary resuscitation [6].

FIGURE 1.15   Illustration of some examples of PKMs for medical applications.

[75], a delta-like manipulator was proposed for cardiopulmonary resuscitation (CPR), making chest compression in the patient [6]. The mentioned examples for medical applications are illustrated in Fig. 1.15.

1.2.2.5 Agriculture applications

The agriculture industry is an important area where robotics is making headway due to the hard work that cultivating and preserving crops represents. Such operations make use of chemical or mechanical processes to destroy the weeds [89], [46]. This invention helps considerably to reduce the use of herbicides and fertilizers, which can harm when used in excess. Recently, in the literature it has been reported the design of a robotic system based on PKM designed to pluck the tea. Fig. 1.16 depicts an illustration of the plucking manipulator.

1.2.2.6 Motion simulators

Motion simulators are typical examples of the most relevant applications developed by PKMs. For this purpose, devices must require 6-DOF motions, enough capacity to support heavy loads, fast and accurate movements, and high stiffness. For all these reasons, most of them are based on the Gough-Stewart's and Cappel's platform architecture. In aerospace industry, they play an essential role in aircrafts' design to prevent fatal accidents [4]. In addition to being used as flight simulators, motion simulators have also been used as car, ship, and space simulators [80]. Fig. 1.17 presents illustrates the Sukhoi SuperJet full flight simulator.

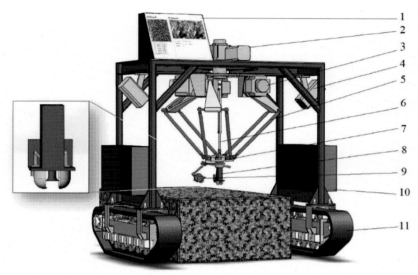

1. Computer 2. Plucked tea shoots collector 3. Servo motor 4. Global camera 5. Delta manipulator 6. Tea shoots suction tube 7. Moving platform 8. End-effector 9. Local camera 10. Power unit 11. caterpillar track mechanism

FIGURE 1.16 A system based on a delta-like manipulator used for agricultural applications [135].

FIGURE 1.17 Sukhoi SuperJet full flight simulator [131].

1.2.2.7 3D printers

Three-dimensional (3D) printing, also known as Additive Manufacturing (AM), is the construction of a three-dimensional object from a

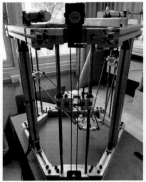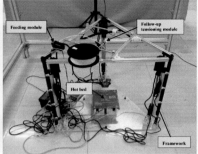

(a) General overview of a 3D printer based on a linear Delta PKM [26].

(b) A detailed view of a linear 3D printer [140].

FIGURE 1.18   View of two examples of 3D printers based on delta-like PKMs.

Computer-Aided Design (CAD) model. These devices build products by adding the material layer by layer rather than by subtracting material from a piece of material as machining tools. In recent years, 3D printers have gained an increased popularity since they manufacture low-cost pieces reducing wasted material. Nowadays, this technology is used to manufacturing automobile components, aircraft components, custom hearing aids, custom orthodontics, among others [23]. The Delta robot with linear actuators is widely used for 3D printing due to the parallelogram bars keeping the traveling plate perfectly horizontal, and the linear actuators provide more rigidity resulting in useful quality end-products [23], [26]. Fig. 1.18 presents some prototypes of 3D printers based on the linear Delta PKM.

1.2.2.8 Haptic devices

Haptics is the science that deals with the study of the sense of touch. Haptic devices allow the user to feel and interact indirectly with an external environment through physical manipulation of the device. These instruments are mainly present in virtual and augmented reality, gaming controllers, and remotely operated surgery. According to [17], some mechanical features, such as low friction, low inertia, high dynamic range, and high stiffness, are essential for high performance in haptic devices. PKMs are well-known for having those mentioned proprieties, so they are relevant candidates to be used as sophisticated haptic interfaces. Over time, it has been proposed haptic devices based on the Gough-Stewart platform and the Delta PKM [102]. For instance, the sigma7 model of the Swiss company Force dimension, which is a 7-DOF haptic interface is based on a 3-DOF Delta PKM [53]. Fig. 1.19 presents a prototype based on the mentioned Sigma 7 platform.

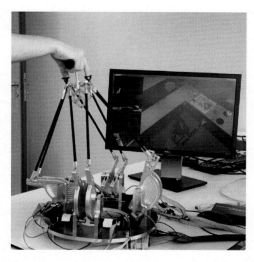

FIGURE 1.19   Illustration of a 7-DOF Haptic device based on a delta-like manipulator [73].

1.3 Control problem formulation

Many factors converge in order for PKMs to successfully develop the previously mentioned applications. These factors range from mechanical design and modeling to trajectory planning and control design. In the literature, it has been reported various works relating to the enhancement of such aspects in order to increase the capabilities of parallel robots [22]. Control design is an important research topic in PKMs since control laws determine the generalized forces/torques that the actuators should generate in order to guarantee the execution of the commanded task while satisfying given transient and steady-state requirements [116]. Following this research line, this book's general purpose is to develop and validate advanced control solutions to improve the overall performance of PKMs under different operating conditions.

1.3.1 PKMs control challenges

Parallel robots are complex nonlinear systems, so controlling them results in a significantly challenging task. From a control point of view, the following considerations should be considered in the design of control schemes for PKMs [31], [29].

1.3.1.1 *Highly nonlinear dynamics*

The closed-kinematic-structure, together with the presence of several passive joints, make PKMs complex nonlinear systems. Their kinematic configuration leads to coupled dynamics, so the actuators' movements should be synchronized with each other not to harm the whole performance of the PKM. According to [86], the effect of nonlinearities in PKMs increases considerably at high-speed/acceleration operation conditions. In this context, high-accelerations lead to mechanical vibrations that may produce mechanical damages and decreasing the precision of the robot. Thus, the use of decentralized linear controllers does not guarantee the safety and stability of PKMs in such operational conditions. For these reasons, advanced nonlinear control techniques should be developed in order to minimize the nonlinearities effects while satisfying the demands of high precision under high-acceleration requirements.

1.3.1.2 *Unstructured and structured uncertainties*

As explained in [107] and [12], uncertainties are the differences between a calculated dynamic model and the real system. They can be classified as structured and unstructured. The first type comprises inaccurate knowledge of the dynamic parameters (e.g., masses and inertias), and dynamic parameters variations as the payload in P&P tasks, or contact forces in machining operations. In contrast, the second type includes geometric manufacturing errors, non-modeled phenomena, dynamic model simplifications (i.e., neglect the actuators' dynamics or friction), sensors' noise, and wear of mechanical elements. Advanced control strategies should then be considered to deal with all these uncertainties ensuring a good dynamic performance and a high precision in the developed task.

1.3.1.3 *Actuation redundancy*

In robotics, actuation redundancy means that the system has more actuators than DOF [81]. Actuation redundancy has been used to improve kinematic capabilities in PKMs, adding one more kinematic chain with its respective actuator. This configuration presents the following advantages:

- Enhanced dexterity,
- Singularity avoidance in the operational workspace,
- Collision avoidance,
- Backlash avoidance,
- Better load distribution reducing the power consumption of the actuators,
- Improved stiffness and, therefore, potentially better precision.

However, actuation redundancy represents a significant challenge in the control of PKMs, since it leads to the generation of internal forces that may create pre-stress in the mechanism without operational motions and can damage the robot's mechanical structure. According to [67], the internal forces can be produced by non-synchronized independent control of the actuators, geometric imperfections, and measurement errors. In the

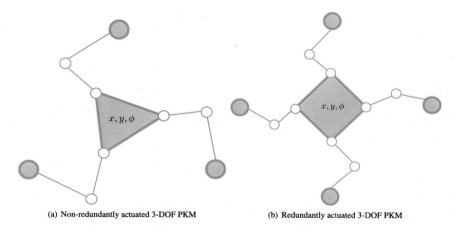

(a) Non-redundantly actuated 3-DOF PKM

(b) Redundantly actuated 3-DOF PKM

FIGURE 1.20   Illustration of the concept of actuation redundancy in PKMs.

literature, actuation redundancy solutions based on the pseudo-inverse Jacobian matrix have been proposed to eliminate those antagonistic internal forces. The concept of actuation redundancy is illustrated in Fig. 1.20 through two 3-DOF PKMs.

2

Literature review about modeling and control of PKMs

2.1 Introduction

As stated in Chapter 1, this book proposes new advanced control solutions for PKMs to fulfill the requirements needed to execute a trajectory tracking task satisfactorily. The proposed controllers should guarantee high precision under different operating conditions despite the presence of uncertainties and external disturbances that may affect the system performance. To test the proposed control techniques, it is necessary to have knowledge about the kinematic and dynamic models of the experimental platforms that will be used. For the case of the Delta PKM, its both kinematic and dynamic models have been well reported in the literature. However, SPIDER4 RA-PKM is a new prototype, and therefore it is necessary to calculate both its kinematic and dynamic models before proceeding with the design and implementation of control schemes. In this way, we establish the other main objective of this book, computing the kinematic and dynamic models of SPIDER4 RA-PKM. Therefore, let us introduce an overview of the dynamic formulations for PKMs before describing the main control schemes reported in the literature.

2.2 Dynamic modeling of parallel kinematic machines

Dynamic modeling formulation is a crucial part in the study of parallel robots, since it allows us to analyze the behavior of the robot under different operating conditions through numerical simulations. The results obtained from simulations may help researchers to propose improvements to PKMs in terms of mechanical design or control to obtain the expected

21

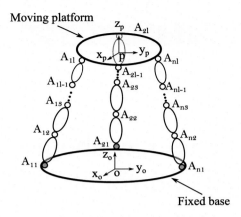

FIGURE 2.1   Illustration of the general kinematic representation of a PKM.

performance under specific operating requirements. Moreover, some advanced control techniques require full or partial knowledge about the
dynamic equations of the PKM to be controlled to improve its tracking
performance. Dynamic analysis of PKMs is complicated due to the closed-
loop structure, involving several kinematic chains. In the literature, the
most common ways to develop dynamic motion equations for a PKM are
based on the Newton-Euler, the principle of virtual works, and Lagrangian
formulation.

2.2.1 Dynamic modeling approaches for PKMs

Before proceeding with explaining of the different formulations to obtain the equations of motion of a parallel robot, let us define to consider
Fig. 2.1 to define the following:

- The fixed base of the PKM is attached to the fixed reference frame $O - x_o, y_o, z_o$.
- A reference frame is located at the moving platform denoted by $p - x_p, y_p, z_p$ whose coordinates are in function of $O - x_o, y_o, z_o$.
- For any parallel robot, the number of DOF will be denoted by m.
- The joints (prismatic, revolute, universal, etc.) are located at the points A_{ij}, where the kinematic chain number is denoted by $i = 1, ..., n$, and the joint number by $j = 1, ..., l$.

2.2.1.1 Newton-Euler formulation

Newton-Euler (NE) is one of the most common formulations used to
compute the dynamic model of PKMs. This methodology is based on the
computation of the motion equations for each body element of the robot.
Applying this formulation is necessary to derive linear and angular accel-

erations of links as well as the moving platform. However, it may lead to a large number of equations resulting in poor computation efficiency [126]. Considering the diagram presented in Fig. 2.1, we define the NE equations for the link set of a PKM that may be expressed in the following form [21]:

$$\sum \mathbf{f}_{ij} = m_{ij}^c \ddot{\mathbf{x}}_{ij}$$
$$\sum {}^c\mathbf{n}_{ij} = {}^c\mathbf{I}_{ij}\dot{\boldsymbol{\omega}}_{ij} + \boldsymbol{\omega}_{ij} \times ({}^c\mathbf{I}_{ij}\boldsymbol{\omega}_{ij})$$

$$(2.1)$$

- m_{ij} is the mass of each ij link,
- $\sum \mathbf{f}_{ij}$ is the sum of external forces acting on each link,
- $\sum {}^c\mathbf{n}_{ij}$ is the sum of external moments including gravity effects, acting of the body center of mass,
- ${}^c\ddot{\mathbf{x}}_{ij}$ is the acceleration to the center of mass of the ij link,
- $\dot{\boldsymbol{\omega}}_{ij}$, $\boldsymbol{\omega}_{ij}$, are the rotational acceleration, and velocity of the ij link,
- ${}^c\mathbf{I}_{ij}$ is the inertia matrix of the ij link expressed at its center of mass.

The NE equations describing the dynamics of the moving platform can be expressed as:

$$\mathbf{f}_p = m_p^c \ddot{\mathbf{x}}_p$$
$${}^c\mathbf{n}_p = {}^c\mathbf{I}_p\dot{\boldsymbol{\omega}}_i + \boldsymbol{\omega}_i \times ({}^c\mathbf{I}_p\boldsymbol{\omega}_p)$$

$$(2.2)$$

In which:

- m_p is the mass of the moving platform,
- ${}^c\mathbf{I}_p$ is the inertia matrix of the moving platform, expressed at its center of mass,
- $\ddot{\mathbf{x}}_p$, and $\dot{\boldsymbol{\omega}}_p$ are the linear and rotational accelerations of the moving platform.

One can mention some literature works where NE formulation has been used to compute Parallel robots' dynamics. For instance, in [19], the Newton-Euler equations were employed to obtain the dynamic model of a 3-DOF Delta PKM. Also, in [63], the Newton-Euler method was applied to get the inverse dynamic model of a Stewart platform.

2.2.1.2 Principle of virtual works

This principle establishes that the virtual work produced by external forces and moments corresponding to any set of virtual displacement of a body or system of bodies is zero. Unlike the NE formulation, this principle does not imply the unknown constraint forces or moments, resulting in less computational efforts. The principle of virtual works applied to a system of N bodies results in the following equation [41]:

$$\delta W = \sum (m_i \ddot{\mathbf{x}}_i - \mathbf{f}_i) \cdot \delta \mathbf{x}_i + \sum ({}^c\mathbf{I}_i\dot{\boldsymbol{\omega}}_i + \boldsymbol{\omega}_i \times ({}^c\mathbf{I}_i\boldsymbol{\omega}_i) - {}^c\mathbf{n}_i) \cdot \delta \boldsymbol{\theta}_i = 0 \quad (2.3)$$

Where δx_i, and $\delta\theta_i$ represent the virtual linear and rotational displacements of the body i, respectively. The previous equation can be adapted to a PKM composed by n links and one moving platform as follows [121]:

$$\delta W = \delta q^T \Gamma + \delta X_p^T \hat{F}_p + \sum \delta X_i^T \hat{F}_i = 0 \tag{2.4}$$

Where:

- δX_p, and δX_i represents the virtual displacements of the moving platform and limbs center of masses, respectively,
- δq denotes the virtual displacements of the joint variables,
- Γ is the vector of the actuator forces,
- $\hat{F}_p$ and $\hat{F}_i$ represent the difference between the generalized inertial forces/moments, and the generalized external forces/moments acting on the moving platform center of mass, and each limb center of mass, respectively, whose components are the following:

$$\hat{F}_p = \begin{bmatrix} m_p \ddot{x}_p - f_p \\ I_p \dot{\omega}_p + \omega_p \times (I_p \omega_p) - n_p \end{bmatrix} \qquad \hat{F}_i = \begin{bmatrix} m_i^c \ddot{x}_i - f_i \\ {}^c I_i \dot{\omega}_i + \omega_i \times ({}^c I_i \omega_i) - {}^c n_i \end{bmatrix}$$

In which m_p denotes the moving platform mass, I_p is the inertia moment of the moving platform, and f_p and n_p represent the external forces and moments acting of the moving platform. The virtual displacements of the joint variables and the limbs are related to the virtual displacement of the moving platform by means of the following relations:

$$\delta q = J \delta X_p \qquad \delta X_i = J_i \delta X_p \tag{2.5}$$

Where J is the Jacobian matrix of the manipulator, and J_i is the limb Jacobian matrix. Substituting (2.5) in (2.4) yields:

$$\delta X_p^T (J^T \Gamma + \hat{F}_p + \sum J_i^T \hat{F}_i) = 0 \tag{2.6}$$

Finally, canceling δX_p^T from the previous expression and rearranging terms, we can express the dynamic model of a PKMs as follows:

$$\Gamma = -J^{-T} (\hat{F}_p + \sum J_i^T \hat{F}_i) \tag{2.7}$$

The virtual work formulation has been extensively used in the dynamic modeling of PKMs, as can be seen in [40], [138], [137], and [74], among other works.

2.2.1.3 Euler-Lagrange formulation

Euler-Lagrange formulation describes the motion based on the kinetic and potential energy of the whole system. This formulation is suitable for dynamic analysis of multibody systems, where each body's forces and moments interact with each other. The Lagrangian function can be expressed in the following form:

$$\mathcal{L} = K - U \tag{2.8}$$

In which K and U represent the total kinetic and potential energy of the system, respectively. According to [121], the Lagrange motion equations can be divided into two types: first kind and second kind. The most straightforward representation of them is the second kind, which can be expressed as follows:

$$\frac{d}{dt}\left(\frac{\partial \mathcal{L}}{\partial \dot{q}_j}\right) - \frac{\partial \mathcal{L}}{\partial q_j} = Q_j, \quad j = 1, 2, ..., m \tag{2.9}$$

Where q_j is the generalized coordinates, Q_j represents the generalized forces, and m expresses the number of the system's independent generalized coordinates. The second kind of Lagrangian motion equation is suitable for manipulators with open-loop kinematic chains. However, the closed-loop configuration of PKMs introduces a set of constraint equations that must be integrated in the motion equations, leading the first kind of Lagrange motion equations, expressed as follows:

$$\frac{d}{dt}\left(\frac{\partial \mathcal{L}}{\partial \dot{q}_j}\right) - \frac{\partial \mathcal{L}}{\partial q_j} = Q_j + \sum_{i=1}^{k} \lambda_i \frac{\partial h_i}{\partial q_j}, \quad j = 1, 2, ..., n \tag{2.10}$$

Where k is the number of constraints, λ_i is the Lagrange multipliers, n represents the number of redundant coordinates exceeding the number of DOF m by k, that is $n = m + k$, and h_i is the constraint function. In order to compute the generalized forces Q_j, it is necessary to split up (2.10) into two sets [126]. The first one contains the Lagrange multipliers as unknown variables, and the second one is the generalized forces as unknown variables. The first set can be written as:

$$\sum_{i=1}^{k} \lambda_i \frac{\partial h_i}{\partial q_j} = \frac{d}{dt}\left(\frac{\partial \mathcal{L}}{\partial \dot{q}_j}\right) - \frac{\partial \mathcal{L}}{\partial q_j} - \hat{Q}_j, \quad j = 1, 2, ..., k \tag{2.11}$$

Being $\hat{Q}_j$ a generalized external force that is supposed to be known. After fining Lagrange multipliers, Actuators' forces and/or torques can be obtained as follows:

$$Q_j = \frac{d}{dt}\left(\frac{\partial \mathcal{L}}{\partial \dot{q}_j}\right) - \frac{\partial \mathcal{L}}{\partial q_j} - \sum_{i=1}^{k} \lambda_i \frac{\partial h_i}{\partial q_j}, \quad j = k+1, ..., n \tag{2.12}$$

Some related works where Euler-Lagrange formulation has been used to obtain the dynamic model equations in Parallel robots include [1], [92], and [55], among others.

2.2.1.4 Simplification-based modeling method for delta-like PKMs

The computation of a suitable dynamic model for real-time implementation of PKMs represents a significant challenge due to the large number of equations, which can become a problem due to the required computing resources consumed by the computer. Parallel robots with delta-like structure have linkages formed by two elements; the first one is the rear-arm formed by a single piece, and the second one is the forearm, which is composed of two parallelogram bars. Computing the dynamic model of this type of robots using any of the presented methodologies above would lead to multiple real-time computing problems. For example, when using the NE formulation, the forces and moments of all the robot elements must be considered, resulting in a large number of equations that require significant computational time to solve them. Alternatively, if one chooses to use the principle of virtual works, the calculation of the inverse link-Jacobian matrices will represent a laborious computation process. For PKMs based on the Delta architecture, some simplification hypotheses have been proposed in [84] to obtain a suitable dynamic model appropriate for real-time implementation. These simplification hypotheses are as follows:

- The frictional forces dry and viscous of the joints are neglected.
- The rotational inertia of the forearms is neglected. Nevertheless, its mass is divided into two equivalent parts; one part is added to the rear-arm mass, and the other one is joined to the traveling plate mass. This simplification is justified if the forearms' mass is highly smaller than the other components of the robot.

Concerning the second simplification, the ratio between the mass of the forearms and the mass of the rear-arms must be sufficiently small, as it is established on the analysis developed for PKMs with delta-like architecture in [84]. Fig. 2.2 illustrates the concept such model simplifications. The resulting equation of motion considering the above simplifications will have the following form:

$$\boldsymbol{\Gamma}(t) = \boldsymbol{\Gamma}_{act} + \boldsymbol{\Gamma}_p \qquad (2.13)$$

Where $\boldsymbol{\Gamma}_{act} \in \mathbb{R}^n$ involves the forces/torque contributions of the actuators and the rear-arms plus a half mass of the forearms, whereas that $\boldsymbol{\Gamma}_p \in \mathbb{R}^n$ represents the forces/torque contribution of the moving platform plus the other half mass of the forearm. This formulation will be considered in Chapter 3 to develop the dynamic models of our experimental platform.

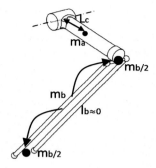

FIGURE 2.2   Illustration of the second modeling simplification for PKMs with delta-like architecture [84].

2.2.2 Dynamic modeling representation

The dynamic modeling of robotic manipulators deals with the relationship between the generalized forces and torques of a robotic system and its corresponding positions, velocities, and accelerations in the Cartesian and joint space. The dynamic modeling can be divided into two branches [118]:

- *Forward dynamics*: It can be defined as follows: given the set generalized forces and torques of the robot, compute the resulting motion of the end-effector and joints as a function of time. The Forward dynamic model expressed in terms of Cartesian space variables can be represented as follows:

$$\ddot{\mathbf{X}} = f(\mathbf{F}, \dot{\mathbf{X}}, \mathbf{X}) \tag{2.14}$$

  where $\mathbf{X}, \dot{\mathbf{X}}, \ddot{\mathbf{X}} \in \mathbb{R}^m$ are the position, velocity, and acceleration of the moving platform, respectively, and $\mathbf{F} \in \mathbb{R}^m$ is the vector of forces/torques applied on the moving platform. The representation of the Forward dynamics in joint space is presented in the following form:

$$\ddot{\mathbf{q}} = f(\boldsymbol{\Gamma}, \dot{\mathbf{q}}, \mathbf{q}) \tag{2.15}$$

  where $\mathbf{q}, \dot{\mathbf{q}}, \ddot{\mathbf{q}} \in \mathbb{R}^n$ are the position, velocity, and acceleration of the actuated joints, respectively, and $\boldsymbol{\Gamma} \in \mathbb{R}^n$ is the vector of actuated joint force/torque.
- *Inverse dynamics*: It can be defined as follows: given a set of trajectories of the end-effector and actuated joints in function of time, find the set of forces/torques that produce this motion. This description may be represented in the following form in terms of Cartesian and joint space coordinates respectively:

$$\mathbf{F} = f(\ddot{\mathbf{X}}, \dot{\mathbf{X}}, \mathbf{X}) \tag{2.16}$$

$$\boldsymbol{\Gamma} = f(\ddot{\mathbf{q}}, \dot{\mathbf{q}}, \mathbf{q}) \tag{2.17}$$

The forces/torques applied to the end-effector and those produced by the actuated joints are related to each other by means of the following expression:

$$\mathbf{F} = \mathbf{J}^T \boldsymbol{\Gamma} \tag{2.18}$$

Where $\mathbf{J} \in \mathbb{R}^{m \times n}$ is the Jacobian matrix of the manipulator. The inverse dynamic modeling plays a crucial role in model-based controllers design. Therefore, we will explain in more details expressions (2.16)–(2.17).

2.2.2.1 Representation of the inverse dynamic model in joint space

The IDM of a robotic system with m DOF and n actuators can be represented for joint space as follows [68]:

$$\mathbf{M}(\mathbf{q})\ddot{\mathbf{q}} + \mathbf{C}(\mathbf{q}, \dot{\mathbf{q}})\dot{\mathbf{q}} + \mathbf{G}(\mathbf{q}) = \boldsymbol{\Gamma}(t) \tag{2.19}$$

Where:

- $\mathbf{q}, \dot{\mathbf{q}}, \ddot{\mathbf{q}} \in \mathbb{R}^n$ are the vectors of position, velocity, and acceleration of the robot in joint space,
- $\boldsymbol{\Gamma} \in \mathbb{R}^n$ is the input torque vector,
- $\mathbf{M}(\mathbf{q}) \in \mathbb{R}^{n \times n}$ is the inertia matrix,
- $\mathbf{C}(\mathbf{q}, \dot{\mathbf{q}}) \in \mathbb{R}^{n \times n}$ is the Coriolis/Centripetal forces matrix,
- $\mathbf{G}(\mathbf{q}) \in \mathbb{R}^n$ is the gravity vector.

This representation is the most common of the IDM for control design purposes because most PKMs do not have sensors to measure the traveling plate's position directly to the fixed coordinates system.

2.2.2.2 Representation of the inverse dynamic model in Cartesian space

The IDM can also be expressed in Cartesian coordinates employing the following relationships based-on the Jacobian matrix.

$$\dot{\mathbf{q}} = \mathbf{J}\dot{\mathbf{X}}, \qquad \ddot{\mathbf{q}} = \mathbf{J}\dot{\mathbf{X}} + \dot{\mathbf{J}}\ddot{\mathbf{X}} \tag{2.20}$$

Substituting (2.20) and (2.18) in (2.19), the IDM in Cartesian space can be written as follows [121]:

$$\mathbf{M}_X \ddot{\mathbf{X}} + \mathbf{C}_X \dot{\mathbf{X}} + \mathbf{G}_X = \mathbf{J}^T \boldsymbol{\Gamma} \tag{2.21}$$

Where:

- $\mathbf{X}, \dot{\mathbf{X}}, \ddot{\mathbf{X}} \in \mathbb{R}^m$ are the position, velocity and acceleration vector of the moving platform of the PKM,
- $\mathbf{M}_X = \mathbf{J}^T \mathbf{M}(\mathbf{q})\mathbf{J}$ is the inertia matrix expressed in Cartesian space,

- $\mathbf{C}_X = \mathbf{J}^T \mathbf{M}(\mathbf{q})\dot{\mathbf{J}} + \mathbf{J}^T \mathbf{C}(\mathbf{q}, \dot{\mathbf{q}})\mathbf{J}$ is the Coriolis/Centripetal matrix expressed in Cartesian space,
- $\mathbf{G}_X = \mathbf{J}^T \mathbf{G}(\mathbf{q})$ is the vector of produced forces due to gravity acceleration expressed in Cartesian space.

2.3 Overview of motion controllers for PKMs

In the literature, several control approaches have been proposed for motion control of parallel robots. Many of them have been taken from control schemes of serial robots since they share similarities in the dynamics mathematical representation. Control solutions for robots can be classified into two primary sets: non-model-based and model-based controllers [104]. The first type is the easiest to implement since they do not require any knowledge about the controlled system dynamics. In contrast, the second type of controllers makes use of the whole or some parts of the inverse dynamic model. In most cases, the obtained tracking performance using model-based controllers outcomes those produced by a non-model-based controller [32]. Other possible classifications of control laws of robots include non-adaptive and adaptive controllers [12]. Fig. 2.3 presents a simple classification of control strategies for PKMs with some examples of controllers existing in the literature. Such examples will be briefly described in the forthcoming section of this chapter except for those based on the RISE control, which will be described in further detail in Chapter 4.

2.4 Non-model-based control schemes

As their name indicates, these control schemes do not need knowledge about the system's dynamics to be controlled. All the required information lies in the measurement of the system's states (positions and velocities). Non-model-based controllers can be subdivided into two categories: Non-adaptive and adaptive controllers, which will be discussed in detail in the following sections.

2.4.1 Non-model-based non-adaptive controllers

Implementing these control schemes on parallel robots is generally painless and straightforward since they do not require any prior knowledge about the manipulator dynamics. The data provided from the robot measurements are processed in a feedback control loop, which sends the control commands to the robot actuators. Most of them have fixed feedback gains with relatively enough good performance for not very demanding tasks. However, to increase these controllers' robustness against external disturbances and uncertainties, it has been proposed to replace

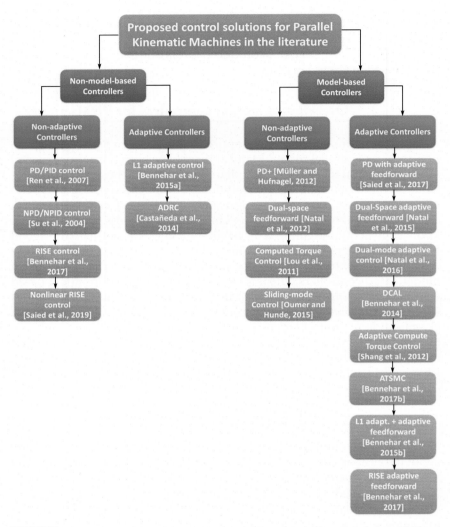

FIGURE 2.3   Classification of proposed control schemes in the literature for parallel kinematic manipulators.

the fixed gains with nonlinear time-varying gains. So, let us start with the description of them.

2.4.1.1 PD/PID controllers

The Proportional-Integral-Derivative (PID) controller [141] is the most used controller in industry, thanks to its simplicity and relatively good performance. However, the performance of this controller may decrease considerably if the system is subjected to high accelerations. In the literature, PID control has been successfully implemented on parallel robots in both joint space [101] and Cartesian space [86]. For joint space, the equation of the control law can be written as:

$$\Gamma(t) = K_p e_q(t) + K_d \frac{d}{dt} e_q(t) + K_i \int_0^t e_q(\tau) d\tau \tag{2.22}$$

Where K_p, K_d, and $K_i \in \mathbb{R}^{n \times n}$ are positive definite diagonal matrices for the proportional, derivative, and integral actions, respectively. The tracking error in joint space $e_q(t) = q_d(t) - q(t)$ is defined as the difference between the desired trajectory in join space $q_d(t) \in \mathbb{R}^n$, and the measured one $q(t) \in \mathbb{R}^n$. Eq. (2.22) represents the most common way to implement a PID controller on PKMs since most of them only dispose of encoders to measure the joint positions of the actuators. Nevertheless, for those PKMs disposing of sensors that allow them to measure the position of the moving platform directly, it is recommended to implement the PID controller in Cartesian space. Its expression is slightly different from the previous one and can be expressed as follows:

$$F(t) = K_p e_X(t) + K_d \frac{d}{dt} e_X(t) + K_i \int_0^t e_X(\tau) d\tau \tag{2.23}$$

Where $e_X(t) = X_d(t) - X(t)$ denotes the tracking error in Cartesian space, being $X_d(t) \in \mathbb{R}^m$ the desired trajectory, and $X(t) \in \mathbb{R}^m$ the measured one, K_p, K_d, and $K_i \in \mathbb{R}^{m \times m}$ are the controller feedback gains. Since (2.23) computes $F(t)$ instead of $\Gamma(t)$, the actuator's control signals must be computed through the Jacobian matrix $J \in \mathbb{R}^{m \times n}$, or the pseudoinverse Jacobian matrix $H \in \mathbb{R}^{m \times n}$ if the PKM is non-redundantly or redundantly actuated respectively. These relationships are indicated in (2.17) and (2.18).

2.4.1.2 Nonlinear PD/PID controllers

Nonlinear PD/PID controllers arise from the need to improve the response and tracking performance of standard PID controllers with fixed gains. The nonlinear time-varying gains endow the controller with a better ability to reject external disturbances. Furthermore, this controller is less sensitive than its counterpart with fixed gains [111]. The mathematical expression of the Nonlinear PID controller applied to a robotic manipulator can be written in joint space as:

$$\Gamma(t) = K_p(.)e_q(t) + K_d(.) \frac{d}{dt} e_q(t) + K_i(.) \int_0^t e_{q(\tau)} d\tau \tag{2.24}$$

Where $K_p(.)$, $K_d(.)$, and $K_i(.) \in \mathbb{R}^{n \times n}$ are the nonlinear time-varying feedback gains for the proportional, derivative, and integral actions, respectively. The controller's gains are automatically tuned according to a series of rules based on the measurement of the tracking error signal. If the tracking error is large, the gains' values will increase their values, making a quick correction action. Once the error decreases, the feedback gains are automatically reduced to avoid excessive oscillations and overshoots. Several methodologies have been proposed in the literature to adjust the nonlinear time-varying feedback gains [88], [3], [111], among others. One can mention an example application in parallel robots. For instance, in [120], a nonlinear PID controller in joint space was proposed to regulate the positioning of a 6-DOF PKM.

2.4.2 Non-model-based-adaptive controllers

Control schemes of this kind do not consider any information on the system dynamics. However, they exploit the information available from the system states to online-estimate nonlinearities, uncertainties, and in some cases, the system dynamic behavior. If the adaptation law is designed to estimate the system's uncertainties and nonlinearities, then the controller will reduce their effects. On the other hand, if the adaptation law is designed to estimate the system's dynamics, this will be integrated as a compensation term in the control scheme. Here are some control schemes that belong to this category:

2.4.2.1 $\mathcal{L}_1$ adaptive control

$\mathcal{L}_1$ adaptive control is a recent control strategy published for the first time in [24] and [25]. This control technique is inspired from the MRAC (Model Reference Adaptive Control) but decoupling the feedback control loop from the adaptive estimation loop. Moreover, this controller guarantees the controller signals to stay in a low-frequency range. In [11], the first implementation of the $\mathcal{L}_1$ adaptive control on a PKM is reported. Two independent terms compose the controller equation, a fixed state-feedback term, characterizing the evolution of the transient response of the PKM and an adaptive term that compensates for the nonlinearities of the system, that is:

$$\Gamma(t) = \mathbf{A}_m \mathbf{r}(t) + \Gamma_{ad}(t) \tag{2.25}$$

Where $\mathbf{A}_m \in \mathbb{R}^{n \times n}$ is a Hurwitz matrix, $\mathbf{r}(t) \in \mathbb{R}^n$ is the combined tracking error in joint space $\mathbf{r}(t) = \dot{\mathbf{e}}_q(t) + \Lambda \mathbf{e}_q(t)$ where $\Lambda \in \mathbb{R}^{n \times n}$ is a positive definite gain matrix. The adaptive control term $\Gamma_{ad}(t)$ is the output of the following equation:

$$\Gamma_{ad}(s) = \mathbf{C}(s)\hat{\boldsymbol{\eta}}(s) \tag{2.26}$$

In which, $\hat{\boldsymbol{\eta}}(s)$ is the Laplace transform of $\hat{\boldsymbol{\eta}}(t) = \hat{\boldsymbol{\theta}}(t)\|\mathbf{r}(t)\|_{\mathcal{L}_\infty} + \hat{\boldsymbol{\varsigma}}(t)$, and $\mathbf{C}(s)$ is a low-pass filter. $\hat{\boldsymbol{\theta}}(t)$, and $\hat{\boldsymbol{\varsigma}}(t)$ are nonlinear estimation functions for the nonlinearities and disturbances of the system, respectively. The procedure to estimate such nonlinear functions is detailed in [12], [13], and [66].

2.4.2.2 *Active disturbance rejection control*

Like the $\mathcal{L}_1$ adaptive control, the Active Disturbance Rejection Control (ADCR) does not require any knowledge about the dynamics of the system to be controlled. This controller on-line estimates nonlinearities and parametric uncertainties for a subsequent proper cancellation of them. The estimation is frequently obtained through Extend State Observers (ESOs) being a kind of high gain observers [54]. To understand the control law's logic in broad strokes, the dynamic model of a robot must be expressed as follows:

$$\ddot{\mathbf{q}} = \mathbf{M}(\mathbf{q})^{-1}\mathbf{\Gamma} - \mathbf{M}(\mathbf{q})^{-1}[\mathbf{C}(\mathbf{q},\dot{\mathbf{q}})\dot{\mathbf{q}} + \mathbf{G}(\mathbf{q})] \tag{2.27}$$

The previous model can be rewritten in state space as follows:

$$\begin{aligned} \dot{\mathbf{x}}_a(t) &= \mathbf{x}_b(t) \\ \dot{\mathbf{x}}_b(t) &= f(\dot{\mathbf{x}}_T(t)) + g(\mathbf{x}_a(t))\mathbf{\Gamma}(t) + \chi(\mathbf{x}_T(t),t) \end{aligned} \tag{2.28}$$

Where $\mathbf{x}_T \in \mathbb{R}^{2n}$ is the robotic system state vector, that is $\mathbf{x}_T = [\mathbf{x}_a \quad \mathbf{x}_b]^T$. Being $\mathbf{x}_a = \mathbf{q}$, and $\mathbf{x}_b = \dot{\mathbf{q}}$, the term $\chi(\mathbf{x}_T(t),t)$ considers the presence of uncertainties and non-modeled phenomena. With the system structure presented in (2.28) the authors proposed to design a high-gain observer resulting in a control law with the following structure:

$$\mathbf{\Gamma}(t) = -g(\mathbf{x}_a)^{-1}(\mathcal{H}^T\hat{\mathbf{x}}_T(t) + \hat{\mathbf{\Lambda}}(t)\hat{\mathbf{x}}_T(t)) \tag{2.29}$$

Where, $\mathcal{H} \in \mathbb{R}^{2n}$ is a linear constant gain matrix, and $\hat{\mathbf{\Lambda}} \in \mathbb{R}^{2n \times n}$ is a matrix containing the time-varying parameters, which are automatically adjusted by an adaptive rule based on the least mean square error. The proposed ADRC was successfully implemented on a 3-DOF Delta PKM in [28].

2.5 Model-based control schemes

Control schemes belonging to this category incorporate complete or partial knowledge of the system's dynamics to compensate the effects of nonlinearities, enhancing the overall performance of the robotic system. In most of the cases reported in the literature, the performance obtained with these control schemes far exceeds the one of non-model-based controllers [104].

2.5.1 Model-based-non-adaptive controllers

As the name of this subcategory indicates, the dynamic parameters of the robot remain fixed, which may have certain advantages and disadvantages. On the one hand, if a precise knowledge of such parameters exists, the performance obtained will be satisfactory. However, if such parameters are poorly known, instead of improving the robot's performance, it will degrade it. Some examples of this type of control schemes include the following:

2.5.1.1 Computed torque control

Computed Torque Control (CTC), also known in some literature as Inverse Dynamics Control (IDC) [121], is a control technique that has been widely applied in the control of robots. It is a model-based non-adaptive control technique that allows us, under some assumptions, to obtain a linear closed-loop equation in terms of the tracking errors. This control technique may provide an excellent tracking performance if the model and the system parameters are precisely known. The control law applied to any manipulator in joint space can be established as follows [2]:

$$\mathbf{\Gamma}(t) = \mathbf{M}(\mathbf{q})(\ddot{\mathbf{q}}_d + \mathbf{K_p}\mathbf{e}_q(t) + \mathbf{K_d}\dot{\mathbf{e}}_q(t)) + \mathbf{C}(\mathbf{q}, \dot{\mathbf{q}})\dot{\mathbf{q}} + \mathbf{G}(\mathbf{q}) \qquad (2.30)$$

where $\mathbf{K_p}, \mathbf{K_d} \in \mathbb{R}^{n \times n}$ are positive definite diagonal feedback gain matrices for the proportional, and derivative control actions, respectively. $\mathbf{e}_q(t) \in \mathbb{R}^n$, and $\dot{\mathbf{e}}_q(t) \in \mathbb{R}^n$ are the tracking errors of the joint positions and velocities, respectively. Substituting this control law in the inverse dynamic model in joint space (2.19), leads to the following closed-loop system dynamics:

$$\ddot{\mathbf{e}}_q(t) + \mathbf{K_d}\dot{\mathbf{e}}_q(t) + \mathbf{K_p}\mathbf{e}_q(t) = \mathbf{0} \qquad (2.31)$$

In which $\ddot{\mathbf{e}}_q$ represents the acceleration error in joint space. It is worth noting that the previous equation is valid under the assumption that all the dynamic parameters are entirely known. Moreover, as it can be seen in (2.28), if the controller gains are chosen appropriately, the tracking error can converge quickly towards zero with a suitable steady-state performance. However, this control technique is computationally intensive, and the use of the measured variables as joint positions and velocities may yield noise that may harm the system performance. In [78], the CTC scheme was used to control the positioning of a 3-DOF PKM called Orthopod.

2.5.1.2 Augmented PD control

This control scheme is also known as PD+. It is composed of a PD feedback control term plus the inverse dynamic model of the robot based on the desired and the measured trajectories. One key difference with the CTC scheme is that the inertia matrix is outside the position and velocity feedback loops yielding a faster computation process [90]. The control law is relatively simple, and can be expressed as:

$$\mathbf{\Gamma}(t) = \mathbf{K_p}\mathbf{e}_q(t) + \mathbf{K_d}\dot{\mathbf{e}}_q(t) + \mathbf{M}(\mathbf{q})(\ddot{\mathbf{q}}_\mathbf{d}) + \mathbf{C}(\mathbf{q}, \dot{\mathbf{q}})\dot{\mathbf{q}}_\mathbf{d} + \mathbf{G}(\mathbf{q}) \qquad (2.32)$$

Where $\mathbf{K_p}, \mathbf{K_d} \in \mathbb{R}^{n\times n}$ are positive definite diagonal feedback gain matrices, for the proportional, and derivative control actions respectively. $\mathbf{e}_q(t) \in \mathbb{R}^n$, and $\dot{\mathbf{e}}_q(t) \in \mathbb{R}^n$ are the tracking errors of the joint positions and velocities, respectively. The feedback PD control term guarantees asymptotic stability, and the model-based part cancels the effect of some nonlinearities. However, to this end, the dynamic parameters used in the equation must be relatively close to the real ones. The original PD+ has been modified in different ways to improve its performance. For instance, in [82], a PD+ controller in redundant coordinates was proposed as an alternative to coordinate switching and applied to the control of redundantly actuated PKMs.

2.5.1.3 PD computed feedforward control

This controller consists of a PD controller plus nominal feedforward terms. Its expression is mainly similar to the previous controller. However, the whole inverse dynamic model is evaluated with the desired trajectories instead of the measured ones since the sensors used to measure the actual joint variables may integrate noise in the measurements, which can impair the robot performance. PD feedforward is one of the most used control techniques in robotic due to its simplicity, and its asymptotic stability. The expression of its control law is given as follows [110]:

$$\mathbf{\Gamma}(t) = \mathbf{K_p}\mathbf{e}_q(t) + \mathbf{K_d}\dot{\mathbf{e}}_q(t) + \mathbf{M}(\mathbf{q_d})\ddot{\mathbf{q}}_\mathbf{d} + \mathbf{C}(\mathbf{q_d}, \dot{\mathbf{q}}_\mathbf{d})\dot{\mathbf{q}}_\mathbf{d} + \mathbf{G}(\mathbf{q_d}) \qquad (2.33)$$

where $\mathbf{K_p}, \mathbf{K_d} \in \mathbb{R}^{n\times n}$ are positive definite diagonal feedback gain matrices for the proportional, and derivative control actions respectively. $\mathbf{e}_q(t) \in \mathbb{R}^n$, and $\dot{\mathbf{e}}_q(t) \in \mathbb{R}^n$ are the tracking errors of the joint positions and velocities, respectively. In [86], a modified version of the standard PD feedforward named "Dual-Space Control" was validated experimentally in a RA-PKM called R4. The difference of this control scheme w.r.t. the standard PD feedforward is that the dynamic model part is formulated in both coordinate spaces, Cartesian and joint and not only in one as with the original formulation.

2.5.1.4 Higher order sliding mode control

Sliding Mode Controllers (SMCs) are robust nonlinear control techniques appropriate for nonlinear uncertain systems. The main feature of those control schemes lies in the use of a discontinuous function with a high control gain leading to finite-time convergence and robustness of the closed-loop system. However, these controllers have a drawback regarding the produced undesirable high-frequency oscillations called chattering

[128]. This unwanted phenomenon is very dangerous for the system actuators. Therefore, to overcome that issue, Higher-Order Sliding Mode Controllers (HOSMCs) have been proposed. Many methodologies based on HOSMCs have been proposed for the control of PKMs. For instance, in [115], a trajectory tracking control based on a HOSMC was implemented on a 2-DOF PKM driven by pneumatic muscle actuators. In [71], a HOSMC was proposed for the stabilization of a Stewart-platform. The velocities of such robots were estimated through an observer based on the super-twisting algorithm. Numerical simulations were performed to show the effectiveness of the proposed controller.

2.5.2 Model-based-adaptive controllers

Model-adaptive-based controllers emerge because, in reality, the dynamic parameters of a manipulator (as masses or inertias) are challenging to know with precision, and most of the time, some of such parameters may vary according to the assigned task. Many of the control schemes belonging to this subcategory are improvements of model-based controllers with fixed parameters. Therefore, the dynamic model representation must be rewritten in a parameterized way, decoupling the dynamic parameters from known linear and nonlinear functions of the dynamic model. Considering (2.19), the IDM can be reformulated as:

$$\mathbf{M}(\mathbf{q})\ddot{\mathbf{q}} + \mathbf{C}(\mathbf{q}, \dot{\mathbf{q}})\dot{\mathbf{q}} + \mathbf{G}(\mathbf{q}) = \mathbf{Y}(\mathbf{q}, \dot{\mathbf{q}}, \ddot{\mathbf{q}})\mathbf{\Phi} \qquad (2.34)$$

In which $\mathbf{Y}(\mathbf{q}, \dot{\mathbf{q}}, \ddot{\mathbf{q}}) \in \mathbb{R}^{n \times r}$ is known as the regressor matrix involving all linear and nonlinear functions of the model dynamics, and $\mathbf{\Phi} \in \mathbb{R}^{r}$ is the vector parameters, where the superscript r indicates the number of parameters to be estimated. The right side of (2.34) will be considered in the description of the following control solutions.

2.5.2.1 Adaptive computed torque control

This extension of the classical CTC described in the previous section. This modification is due to, in most cases, the dynamic parameters are not well known, they are time-varying, or both. Consequently, the presented nonlinearities of the system will not be entirely canceled by the controller. Therefore, an estimation algorithm should be considered in order to estimate in real-time such unknown parameters. Taking into account (2.30), we can express the closed-loop system for Adaptive Computed Torque Control (ACTC) as [45]:

$$\ddot{\mathbf{e}}_q(t) = -\mathbf{K}_p\mathbf{e}_q(t) - \mathbf{K}_d\dot{\mathbf{e}}_q(t) + \hat{\mathbf{M}}^{-1}(\mathbf{q})\mathbf{Y}(\mathbf{q}, \dot{\mathbf{q}}, \ddot{\mathbf{q}})\mathbf{e}_\Phi \qquad (2.35)$$

In which, $\hat{\mathbf{M}} \in \mathbb{R}^{n \times n}$ is the estimation of the manipulator inertia matrix, $\mathbf{Y}(\mathbf{q}, \dot{\mathbf{q}}, \ddot{\mathbf{q}})$ is the regressor matrix, and $\mathbf{e}_\Phi$ is the vector of estimation errors

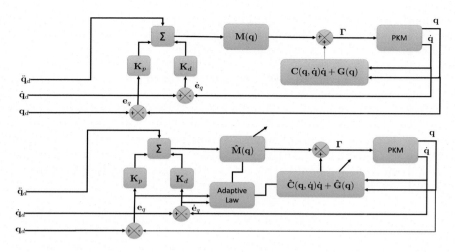

FIGURE 2.4 Block diagram comparison of standard CTC (blue) and ACTC (green).

which is given as $\mathbf{e}_\Phi = \Phi - \hat{\Phi}$ being Φ the parameter vector, and $\hat{\Phi}$ its estimation. The adaptation law is formulated using the following Lyapunov function candidate:

$$V(\boldsymbol{\varepsilon}_q, \mathbf{e}_\Phi) = \boldsymbol{\varepsilon}_q^T \mathbf{P} \boldsymbol{\varepsilon}_q + \mathbf{e}_\Phi^T \boldsymbol{\Xi}^{-1} \mathbf{e}_\Phi \qquad (2.36)$$

Where $\boldsymbol{\varepsilon}_q = \begin{bmatrix} \dot{\mathbf{e}}_q & \mathbf{e}_q \end{bmatrix}^T$, $\mathbf{P}$ is a positive-definite matrix designed to satisfy the Lyapunov equation, and $\boldsymbol{\Xi}$ is a positive-definite constant matrix defined for the adaptation law, which can be expressed as:

$$\dot{\hat{\Phi}} = \boldsymbol{\Xi} \mathbf{Y}^T \hat{\mathbf{M}}^{-1} \mathbf{r}(t) \qquad (2.37)$$

Being $\mathbf{r}(t)$ the combined tracking error in joint space $\mathbf{r}(t) = \dot{\mathbf{e}}_q(t) + \boldsymbol{\Lambda} \mathbf{e}_q(t)$, and $\boldsymbol{\Lambda} \in \mathbb{R}^{n \times n}$ is a positive definite gain matrix. However, like its non-adaptive counterpart, ACTC has the drawback of using measured velocities and accelerations, which are tedious to estimate. This control scheme was applied in [112] to a RA 2-DOF planar parallel robot; the obtained performance was compared to a classical CTC. The results showed a considerable tracking performance improvement of the proposed ACTC compared to CTC. Fig. 2.4 illustrates the control scheme differences between CTC and ACTC.

2.5.2.2 PD with adaptive feedforward

PD control with adaptive feedforward is one of the simplest adaptive-based controllers that can be implemented on robots. Two parts form the controller. On the one hand, the PD feedback control term provides robustness to the system. On the other hand, the adaptive feedforward term intends to cancel nonlinearities and reduce the effect produced by parametric uncertainties. For robotic manipulators, the mathematical expression of this control law can be defined as [65]:

$$\Gamma(t) = \mathbf{K_p}e_q(t) + \mathbf{K_d}\dot{e}_q(t) + \mathbf{Y}(\mathbf{q_d}, \dot{\mathbf{q}}_d, \ddot{\mathbf{q}}_d)\hat{\Phi} \tag{2.38}$$

Where $\mathbf{K_p}$ and $\mathbf{K_d} \in \mathbb{R}^{n \times n}$ are positive-definite diagonal feedback gain matrices for the proportional and derivative control actions, respectively, $\hat{\Phi} \in \mathbb{R}^r$ is the vector of unknown dynamic parameters, being r the number of its elements and $\mathbf{Y}(\mathbf{q_d}, \dot{\mathbf{q}}_d, \ddot{\mathbf{q}}_d) \in \mathbb{R}^{n \times r}$, is the regressor matrix. The unknown dynamic parameters are estimated according to the following adaptation law:

$$\dot{\hat{\Phi}} = \mathbf{K}\mathbf{Y}(\mathbf{q_d}, \dot{\mathbf{q}}_d, \ddot{\mathbf{q}}_d)^T (\mathbf{K_p}e_q(t) + \mathbf{K_d}\dot{e}_q(t)) \tag{2.39}$$

Where $\mathbf{K} \in \mathbb{R}^{r \times r}$ is a positive-definite diagonal matrix that should be carefully adjusted to obtain a good estimation of the parameters while maintaining the closed-loop system stability. This control scheme was successfully implemented in real-time to control the positioning of a 4-DOF PKM called Veloce. The obtained performance was compared to that produced by a PID controller. The results show a superiority of the PD adaptive feedforward controller over the PID controller [104].

2.5.2.3 Desired compensation adaptive control

Desired Compensation Adaptive Control (DCAL) was initially reported in [103]; this model-based control scheme is formed by three parts, a decentralized PD controller, an adaptive feedforward term, and an additional nonlinear compensation term. The control law can be expressed in the following form [14]:

$$\Gamma(t) = \mathbf{K_p}e_q(t) + \mathbf{K_v}r(t) + \mathbf{Y}(\mathbf{q_d}, \dot{\mathbf{q}}_d, \ddot{\mathbf{q}}_d)\hat{\Phi} + \vartheta \|e_q(t)\|^2 r(t) \tag{2.40}$$

Where $\mathbf{K_p}, \mathbf{K_v} \in \mathbb{R}^{n \times n}$ are positive definite diagonal feedback gain matrices, $r(t) \in \mathbb{R}^n$ is the combined tracking error in joint space formulated as $r(t) = \dot{e}_q(t) + \Lambda e_q(t)$. The following expression can formulate the parameter estimation vector:

$$\dot{\hat{\Phi}} = \mathbf{K}\mathbf{Y}^T(\mathbf{q_d}, \dot{\mathbf{q}}_d, \ddot{\mathbf{q}}_d)r(t) \tag{2.41}$$

Where $\mathbf{K} \in \mathbb{R}^{n \times n}$ is a positive-definite adaptation gain matrix. One can highlight the following advantages of the DCAL control scheme: Reduced computing time thanks to using the desired joint trajectories instead of the measured ones in the regressor matrix and the parameters adaptation rule, noise effect reduction, and compensation for parametric uncertainties and nonlinearities. In [8], DCAL control was applied to a RA-PKM called

Dual-V. In the same work, the fixed gains of the feedback controller part were substituted by time-varying nonlinear gains in order to improve the parallel robot's overall performance.

2.5.2.4 Dual-mode adaptive control

This control scheme was implemented on Par2 PKM; it consists of three terms: an adaptive feedforward term, a smooth variable structure term, and a stabilizing term. Dual-mode adaptive control law can be expressed as follows [87]:

$$\mathbf{\Gamma}(t) = \mathbf{K}\mathbf{r}(t) + \overline{d}\frac{\alpha_s \mathbf{r}(t)}{||\alpha_s \mathbf{r}(t)|| + 1} + \mathbf{Y}(\mathbf{q_d}, \dot{\mathbf{q}}_d, \ddot{\mathbf{q}}_d)\hat{\mathbf{a}} \qquad (2.42)$$

Where $\mathbf{K} \in \mathbb{R}^{n \times n}$ is a positive-definite gain matrix, α, and $\overline{d}$ are positive constants. $\hat{\mathbf{a}}$ represents an estimation of the unknown parameters of the system $\hat{\mathbf{a}}_{nom}$. The adaptation rule is given as follows:

$$\dot{\hat{\mathbf{\Phi}}} = -\gamma \text{Proj}(\mathbf{r}(t)^T \mathbf{Y}, \hat{\mathbf{\Phi}}) = -\sigma_r \hat{\mathbf{\Phi}} - \gamma \mathbf{r}(t)^T \qquad (2.43)$$

Where, $\hat{\mathbf{\Phi}}$ is the difference between the currently estimated values and the nominal values of the parameters, γ is the adaptation constant gain, and σ_r is a variable expressed as follows:

$$\sigma_r = \begin{cases} 0 & \text{if } ||\hat{\mathbf{\Phi}}|| < M_\Phi \quad \text{or} \quad \sigma_{eq} < 0 \\ \sigma_{eq} & \text{if } ||\hat{\mathbf{\Phi}}|| \geq M_\Phi \quad \text{or} \quad \sigma_{eq} \geq 0 \end{cases} \qquad (2.44)$$

$$\sigma_{eq} = -\frac{\gamma \mathbf{r}(t)^T \mathbf{Y}\hat{\mathbf{\Phi}}}{||\hat{\mathbf{\Phi}}||^2} \qquad (2.45)$$

Where M_Φ is the maximum possible value of the inertia matrix, and $||\hat{\mathbf{\Phi}}||^2$ is the 2-norm of $\hat{\mathbf{\Phi}}$. According to [85], the adaptation law of Dual-Mode adaptive control provides some advantages w.r.t. other robust control algorithms as the generation of continuous control signals, improvement in the system robustness, less sensitivity to measurement noises. Moreover, for large tracking errors, the controller behaves like a sliding mode controller generating exponential convergence to a residual domain arbitrarily small, and to smaller errors, it behaves as a parametric adaptation law.

2.5.2.5 $\mathcal{L}_1$ adaptive control with adaptive feedforward

Although the original $\mathcal{L}_1$ adaptive controller does not require knowledge of the system's dynamics, there are still nonlinearities and uncertainties in which the adaptive compensation term $\mathbf{\Gamma}_{ad}(t)$ may not been able to eliminate. In order to improve the capabilities of the original $\mathcal{L}_1$ adaptive controller, in [10], it was proposed to integrate a model-based adaptive

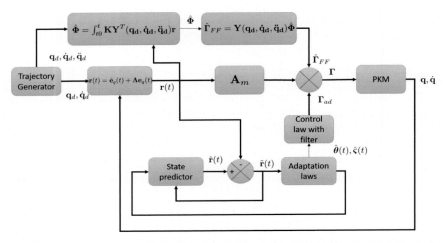

FIGURE 2.5 Block diagram of $\mathcal{L}_1$ adaptive control for PKM (blue), plus adaptive feedforward compensation (green).

feedforward term to reduce further the effect of such nonlinearities and uncertainties of the system. Thus, the control law (2.25) is extended as follows:

$$\mathbf{\Gamma}(t) = \mathbf{A}_m \mathbf{r}(t) + \mathbf{\Gamma}_{ad}(t) + \mathbf{Y}(\mathbf{q_d}, \dot{\mathbf{q}}_d, \ddot{\mathbf{q}}_d)\hat{\mathbf{\Phi}} \qquad (2.46)$$

Where $\mathbf{Y}(\mathbf{q_d}, \dot{\mathbf{q}}_d, \ddot{\mathbf{q}}_d) \in \mathbb{R}^{n \times r}$, is the regressor matrix, and $\hat{\mathbf{\Phi}} \in \mathbb{R}^r$, is the vector of estimated dynamic parameters whose elements are computed using the adaptation rule presented in (2.37). This control law was successfully validated through real-time experiments on the RA 4-DOF Arrow PKM. The obtained results show an improvement in the trajectory tracking performance of more than 78% compared to the original $\mathcal{L}_1$ adaptive controller. Fig. 2.5 presents the control diagram of the standard $\mathcal{L}_1$ adaptive controller in blue color plus the adaptive-model-based term proposed in [10] in green color.

2.5.2.6 Adaptive terminal sliding mode control

Terminal Sliding Mode Controllers (TSMCs) are a kind of HOSMCs with finite-time convergence, maintaining the robustness of conventional SMCs. The first designs of TSMCs had a singularity issue stemming from the sliding manifold's design [129]. Nevertheless, in some research works, different solutions have been proposed to deal with such a problem redesigning the sliding manifold [52]. Considering the advantages provided by TSMCs and adaptive controllers. In [15], a novel adaptive TSMC was proposed for motion control of PKMs, where an adaptation loop on-line estimates the unknown, uncertain, and time-varying parameters of the model-based controller. The proposed control law can be expressed as follows:

$$\mathbf{\Gamma}(t) = \mathbf{Y}(\mathbf{q_d}, \dot{\mathbf{q}}_{\mathbf{d}}, \ddot{\mathbf{q}}_{\mathbf{d}})\hat{\mathbf{\Phi}} - \mathbf{K}_1\mathbf{s} - \mathbf{K}_2|\mathbf{s}|^{\rho}\mathrm{sgn}(\mathbf{s}) \qquad (2.47)$$

In which, $\mathbf{Y}(\mathbf{q_d}, \dot{\mathbf{q}}_{\mathbf{d}}, \ddot{\mathbf{q}}_{\mathbf{d}}) \in \mathbb{R}^{n \times r}$, is the regressor matrix evaluated with the desired joint trajectories, and $\hat{\mathbf{\Phi}} \in \mathbb{R}^r$, is the vector of estimated dynamic parameters, $\mathbf{K}_1$, $\mathbf{K}_2 \in \mathbb{R}^{n \times n}$ are positive-definite diagonal feedback gain matrices, and ρ is an auxiliary control term. The sliding manifold is designed as $\mathbf{s} = \mathbf{e}_q + \beta|\dot{\mathbf{e}}_q|^{\gamma}\mathrm{sgn}(\dot{\mathbf{e}}_q)$, being $0 < \gamma < 2$ and $\beta > 0$ control design parameters. The unknown parameters are estimated according to the following adaptation rule:

$$\hat{\mathbf{\Phi}} = -\mathbf{\Xi}(\mathbf{q_d}, \dot{\mathbf{q}}_{\mathbf{d}}, \ddot{\mathbf{q}}_{\mathbf{d}})^T (\mathbf{K}_3\mathbf{s} + \mathbf{K}_4|\mathbf{s}|^{\rho}\mathrm{sgn}(\mathbf{s})) \qquad (2.48)$$

Where $\mathbf{\Xi} \in \mathbb{R}^{r \times r}$ is a positive-definite diagonal feedback gain matrix known as the adaptation gain matrix, and $\mathbf{K}_3$, $\mathbf{K}_4 \in \mathbb{R}^{n \times n}$ are design matrices for the adaptation law. This proposed adaptive TSMC was validated through real-time experiments on a PKM called Veloce in [15]. The obtained tracking performance was compared to the one produced by nominal TSMC, obtaining more than 50% improvement.

Description and modeling of experimental platforms

3.1 Introduction

Modeling is a crucial aspect in the study of any robotic system since it allows to describe the mechanism's behavior and how the different mechanical parameters may affect the system's overall performance. Additionally, it helps in the design of control schemes to precisely command the robot movements. The kinematic analysis studies the geometry of a robot's motion, without considering the forces or torques that produced the motion [121]. Most kinematic models in PKMs have developed form geometric analysis of the kinematic chains resulting in an expression involving the joint and Cartesian variables called the closed-loop equation [80]. Obtaining the IKM for a PKM is a relatively easy and straightforward task, unlike serial manipulators. However, obtaining the FKM is a complex task for the majority of PKMs because it involves highly coupled nonlinear equations with a non unique solution. The aim of this chapter is the development of the different kinematic and dynamic models for two PKMs. The first one is a RA-PKM called SPIDER4, and the second is a non redundant PKM (the Delta robot). As stated in Chapter 1, one of the main objectives of the thesis is the modeling of SPIDER4, so we will first focus on developing the kinematic and dynamic models of this prototype, then we will present the modeling equations of the 3-DOF Delta PKM.

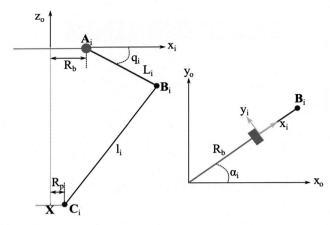

FIGURE 3.1 Side view (at left), and top view (at right) of the kinematic structure of a general delta-like PKM.

3.2 Kinematic modeling for delta-like PKMs

3.2.1 Inverse kinematic formulation

The Inverse Kinematic Model (IKM) consists in finding the position of the final element measured in the fixed reference frame $O - x_o, y_o, z_o$ by given the positions of the angles of the actuators located on the fixed base $\mathbf{q} = \begin{bmatrix} q_1, & \ldots & q_n \end{bmatrix}^T \in \mathbb{R}^{n \times 1}$. For a general delta-like PKM, its kinematic structure can be represented as Fig. 3.1 illustrates, where

- L_i represents the length of each rear-arm
- l_i represents the length of each forearm usually composed by two parallel bars
- R_b is the radius of the fixed base
- R_p is the radius of the traveling-plate
- $\mathbf{A}_i$ is the location vector of each actuator placed in the fixed base
- $\mathbf{B}_i$ is the location vector of each passive joint connecting the rear-arm with the forearm
- $\mathbf{C}_i$ is the location vector of each passive joint connecting the forearm with the traveling-plate
- α_i represents the orientation of each kinematic chain with respect to x_0 axis
- q_i is the angular displacement of each actuator placed in the fixed base
- $\mathbf{X}$ is the vector defining the position of the traveling-plate

The IKM model for delta-like PKMs can be obtained by solving the *closed-loop* equation, which relates the operational space variables $\mathbf{X} \in \mathbb{R}^{m \times 1}$ with the joint variables $\mathbf{q} \in \mathbb{R}^{n \times 1}$

$$\|\mathbf{B}_i \mathbf{C}_i\|^2 = l_i \tag{3.1}$$

In the literature, (3.1) is solved by two methods. The first one was proposed by Calvel/Pierrot in [38], and [95]; the second one way was proposed by Codourey [39]. In this book, the examples will be developed using the methodology proposed by Calvel/Pierrot.

3.2.2 Forward kinematic formulation

Usually for serial robots the computation of the Forward Kinematic Model (FKM) results easy and straightforward. However, for PKMs the computation of the FKM is often more complex yielding multiple solutions [79]. If we consider using (3.1) expression, which has been proposed to solve the IKM problem, when clearing the variable of interest (in this case $\mathbf{X} \in \mathbb{R}^{m \times 1}$), we will obtain multiple coupled nonlinear coupled algebraic equations which will be complicated to solve analytically. In contrast to other parallel robots, the traveling plate of delta-like PKMs always remains oriented in the same plane thanks to the kinematic restrictions imposed by the parallelogram-based structure that composes each forearm. This last data is relevant, because one can calculate the FKM by using a simpler geometric analysis assuming that the moving platform will always have the same orientation. For delta-like PKMs, the most widely used analytically formulation is called the *Virtual-spheres-intersection approach* [132], [61]. Each virtual-sphere is set with a radius l_i whose center by translating horizontally the coordinates of point $\mathbf{B}_i \in \mathbb{R}^{n \times 1}$ by a distance R_p in the direction towards the fixed reference frame. The vector representing the coordinates of the centers of the virtual spheres will be denoted by $\mathbf{B}'_i \in \mathbb{R}^{n \times 1}$. Thus, FKM solution can be found by solving the following equations' system:

$$(x - x_i)^2 + (y - y_i)^2 + (z - z_i)^2 = l_i^2 \quad \forall i = 1, \ldots, n \tag{3.2}$$

Where x_i, y_i, and z_i represents the coordinates of the center of the sphere whose values are directly related to the joint space variables $\mathbf{q} \in \mathbb{R}^{n \times 1}$. It should be noted that in the end two solutions will be obtained because the spheres will intersect at two points. However, the correct solution will be easily identified thanks to a visual analysis as shown in Fig. 3.2.

3.2.3 Jacobian computation

The Jacobian matrix for delta-like manipulators can be formulated from the closed-loop equation described in previous subsections. By developing (3.1) for all kinematic chains of the delta-like manipulator, and concatenating resulting the algebraic expressions, we obtain a nonlinear algebraic

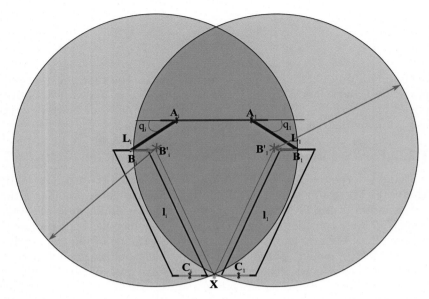

FIGURE 3.2   Side-view illustration of the virtual-spheres-intersection approach used to find the FKM in delta-like PKMs.

function such as:

$$f(\mathbf{X}, \mathbf{q}) = 0 \tag{3.3}$$

where $\mathbf{X} \in \mathbb{R}^{m \times 1}$ contains the Cartesian space variables, and $\mathbf{q} \in \mathbb{R}^{n \times 1}$, the joint space ones. By performing the time derivative of (3.3), leads to [121]:

$$\mathbf{J}_x \dot{\mathbf{X}} = \mathbf{J}_q \dot{\mathbf{q}} \tag{3.4}$$

where $\mathbf{J}_x \in \mathbb{R}^{n \times m}$ and $\mathbf{J}_q \in \mathbb{R}^{n \times n}$ are Jacobian matrices relating the Cartesian and joints space variables, respectively, in which,

$$\mathbf{J}_x = \frac{\partial f}{\partial \mathbf{X}}, \quad \mathbf{J}_q = -\frac{\partial f}{\partial \mathbf{q}} \tag{3.5}$$

Therefore, the general Jacobian matrix of a PKM $\mathbf{J} \in \mathbb{R}^{n \times m}$ is defined as follows:

$$\mathbf{J} = \mathbf{J}_q^{-1} \mathbf{J}_x \tag{3.6}$$

For non-redundant PKMs (i.e. $m = n$), the Jacobian matrix relates the velocities of joint space and Cartesian space as follows:

$$\dot{\mathbf{q}} = \mathbf{J}\dot{\mathbf{X}}, \quad \dot{\mathbf{X}} = \mathbf{J}_m \dot{\mathbf{q}} \tag{3.7}$$

where $\mathbf{J}_m \in \mathbb{R}^{m \times n}$, denotes the inverse of the Jacobian matrix. In the case of redundant PKMs, the velocity relationships of (3.7), will be:

$$\dot{\mathbf{q}} = \mathbf{J}\dot{\mathbf{X}}, \quad \dot{\mathbf{X}} = \mathbf{H}\dot{\mathbf{q}} \tag{3.8}$$

In which $\mathbf{H} \in \mathbb{R}^{m \times n}$ represents the pseudoinverse Jacobian matrix computed as:

$$\mathbf{H} = (\mathbf{J}^T\mathbf{J})^{-1}\mathbf{J}^T \tag{3.9}$$

In addition to relating velocities in Cartesian and joint space, the Jacobian matrix allows mapping the resultant wrenches $\mathcal{F} \in \mathbb{R}^{m \times 1}$ on the traveling plate given the torques/forces produced by the actuators $\mathbf{\Gamma} \in \mathbb{R}^{n \times 1}$ for non-redundant and redundant PKMs as follows:

$$\mathcal{F} = \mathbf{J}^T\mathbf{\Gamma}, \quad \mathbf{\Gamma} = \mathbf{J}_m\mathcal{F} \tag{3.10}$$

$$\mathcal{F} = \mathbf{J}^T\mathbf{\Gamma}, \quad \mathbf{\Gamma} = \mathbf{H}\mathcal{F} \tag{3.11}$$

3.3 Dynamic modeling for delta-like PKMs

3.3.1 Principle of modeling

As explained in Chapter 2, the method for modeling delta-like manipulators was proposed for a suitable implementation in real-time experiments. Now, let us remind the following hypothesis simplification described in Chapter 2.

> The rotational inertia of the forearms is neglected. Nevertheless, its mass is divided into two equivalent parts; one part is added to the rear-arm mass, and the other one is joined to the traveling plate mass (see Fig. 2.2).

This modeling is based on a principle of superposition of the torques applied by each body of the robot, while neglecting the parts whose dynamic effects are weak. Concretely, the inertial effect of a space parallelogram can be neglected under certain conditions (which are explained in Chapter 2), and its mass is represented by two point masses located at each of its ends [95], [84].

Thus, the complete sum of the driving torques of the mechanism (if the actuation is rotary) or forces (if the actuation is linear) can be obtained by the addition of two main contributions: the effect of the actuation system and the effect of the traveling plate, i.e.:

$$\mathbf{\Gamma}(t) = \mathbf{\Gamma}_{sys}(t) + \mathbf{\Gamma}_{tp}(t) \tag{3.12}$$

The components of (3.12) are described in the following subsections.

3.3.2 Torques of forces due to the actuation system

The contribution of the actuation can be written as the sum of several torques (or efforts) due to the following effects:

- **Inertia of the actuator and gearbox**

$$\mathbf{\Gamma}_1 = \mathbf{I}_{act}\ddot{\mathbf{q}} \qquad (3.13)$$

where $\ddot{\mathbf{q}} \in \mathbb{R}^{n \times 1}$, is the vector of accelerations of the joint space variables, $\mathbf{I}_{act}\ddot{\mathbf{q}} = \text{diag}([i_{mot} + i_{gbx}])$, is the $n \times n$ diagonal inertia matrix containing the inertia of the motor i_{mot} and the inertia of the gearbox i_{gbx}.

- **Inertia and torque of the rear-arms**

$$\mathbf{\Gamma}_2 = \mathbf{I}_{ra}\ddot{\mathbf{q}} - \mathbf{M}_{ra}gL_c\cos(\mathbf{q}) \qquad (3.14)$$

where $\mathbf{I}_{ra} = \text{diag}([I_{ra}]) \in \mathbb{R}^{n \times n}$ is a diagonal matrix containing the inertia of the rear-arms, $\cos(\mathbf{q})$ is a vector of $n \times 1$, representing the cosine of each angle q_i, $\forall i = 1...n$, $\mathbf{M}_{ra} = \text{diag}([M_{ra}]) \in \mathbb{R}^{n \times n}$ is the diagonal matrix of the mass of the rear-arms, L_c is the distance from the rotational axis of a rear-arm to its gravity center, and g is the gravity acceleration.

- **Inertia and torque due to the half-mass of the forearm**

$$\mathbf{\Gamma}_3 = \mathbf{I}_{fa}\ddot{\mathbf{q}} - \mathbf{M}_{fa}gL\cos(\mathbf{q}) \qquad (3.15)$$

where L denotes the length of a rear-arm. Considering the second hypothesis simplification, the inertia matrix of the forearms is defined as $\mathbf{I}_{fa} = \text{diag}([L^2(m_{fa})/2]) \in \mathbb{R}^{n \times n}$, the mass matrix of the forearms is then defined as $\mathbf{M}_{fa} = \text{diag}([m_{fa}/2]) \in \mathbb{R}^{n \times n}$, being m_{fa}, the mass of the forearms formed by the spatial parallelograms.

- **Dry and viscous friction of the actuator (motor and gearbox)**

$$\mathbf{\Gamma}_4 = \mathbf{F}_s\text{sgn}(\dot{\mathbf{q}}) + \mathbf{F}_v\dot{\mathbf{q}} \qquad (3.16)$$

where $\mathbf{F}_s\text{diag}([f_s]) \in \mathbb{R}^{n \times n}$ is a diagonal matrix containing the dry friction of the actuators, $\mathbf{F}_v\text{diag}([f_s]) \in \mathbb{R}^{n \times n}$, contains the viscous friction of the actuators, $\dot{\mathbf{q}} \in \mathbb{R}^{n \times 1}$, is the vector containing the speeds of the joint space variables, and $\text{sgn}(\dot{\mathbf{q}})$, represents the vector that gives the sign of such joint speeds.

Therefore, the expression representing the dynamics of the actuation system for a delta-like PKM can be:

$$\mathbf{\Gamma}_{sys} = \mathbf{\Gamma}_1 + \mathbf{\Gamma}_2 + \mathbf{\Gamma}_3 + \mathbf{\Gamma}_4 \qquad (3.17)$$

Or

$$\mathbf{\Gamma}_{sys} = [(\mathbf{I}_{act} + \mathbf{I}_{ra} + \mathbf{I}_{fa})\ddot{\mathbf{q}} - \cos(\mathbf{q})][\mathbf{M}_{ra}gL_c + \mathbf{M}_{fa}gL] + \mathbf{F}_s\text{sgn}(\dot{\mathbf{q}}) + \mathbf{F}_v\dot{\mathbf{q}} \qquad (3.18)$$

3.3.3 Torques due to the traveling plate

In order to determine the torque contributions produced by the traveling plate, we must determine the forces acting on it. These are the gravity and inertia forces.

- **The gravity force on the traveling plate**

$$\mathbf{F}_{tg} = -(\mathbf{M}_{tp} + \mathbf{M}_{nfa})\mathbf{g} \tag{3.19}$$

where $\mathbf{M}_{tp} = \text{diag}([m_{tp}]) \in \mathbb{R}^{n \times n}$, is the diagonal matrix containing the mass of the traveling plate $m_t p$, $\mathbf{M}_{nfa} = \text{diag}([n(m_{fa})/2]) \in \mathbb{R}^{n \times n}$ is the diagonal matrix containing the halves of the mass of the forearms multiplying by n, which represents the number of joint space variables, and $\mathbf{g} = \begin{bmatrix} 0 & 0 & g \end{bmatrix} \in \mathbb{R}^{3 \times 1}$ is the gravity vector.

- **The inertial forces on the traveling plate**

$$\mathbf{F}_{ti} = (\mathbf{M}_{tp} + \mathbf{M}_{nfa})\ddot{\mathbf{X}} \tag{3.20}$$

where $\ddot{\mathbf{X}} \in \mathbb{R}^{3 \times 1}$, represents the acceleration vector of the traveling plate, which is expressed in function if the Cartesian space variables (i.e. x, y, and z).

By relating the forces action on the traveling plate, and using the inverse Jacobian matrix, one can compute its torque contributions as follows:

$$\boldsymbol{\Gamma}_{tp}(t) = \mathbf{J}_m^T(\mathbf{M}_{nfa} + \mathbf{M}_{tp})(\ddot{\mathbf{X}} + \mathbf{g}) \tag{3.21}$$

3.3.4 The general expression

By combining (3.18) and (3.21), the following expression representing the general dynamic model of a delta-like manipulator is obtained.

$$\boldsymbol{\Gamma} = [(\mathbf{I}_{act} + \mathbf{I}_{ra} + \mathbf{I}_{fa})\ddot{\mathbf{q}} - \cos(\mathbf{q})][\mathbf{M}_{ra}gL_c + \mathbf{M}_{fa}gL]$$
$$+ \mathbf{F}_s \text{sgn}(\dot{\mathbf{q}}) + \mathbf{F}_v\dot{\mathbf{q}} + \mathbf{J}_m^T(\mathbf{M}_{nfa} + \mathbf{M}_{tp})(\ddot{\mathbf{X}} + \mathbf{g}) \tag{3.22}$$

This expression will be adapted in further subsections for the 3-DOF Delta robot and SPIDER4. Having presented in a general form the kinematic and dynamic formulations for delta-like manipulators. Let us present the application of the modeling algorithms in the following experimental platforms.

3.4 Application to modeling algorithms to 3-DOF Delta PKM

Delta PKM is a 3-DOF PKM [16], [108] designed to perform high-speed Pick-and-Place (P&P) operations. It has been developed by Reymond Clavel [36] at EPFL-Switzerland. The Delta PKM consists of three

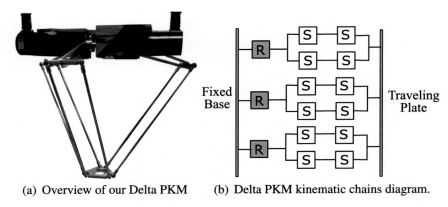

(a) Overview of our Delta PKM   (b) Delta PKM kinematic chains diagram.

FIGURE 3.3   View of Delta PKM and its kinematic configuration.

kinematic chains formed by one rear-arm and one forearm linked together with passive spherical joints. The forearms of the Delta robot are composed of two parallel bars whose purpose is to restrict the orientation of the traveling-plate. The kinematic chains are responsible for the transmission of the movement generated by the three actuators located at the fixed base to the traveling-plate. Fig. 3.3 shows a Delta PKM prototype [49], [47], developed at the Polytechnic University of Tulancingo. The figure also shows its kinematic configuration where the gray boxes represent the active rotational joints, and the white boxes represent the passive spherical joints.

3.4.1 Inverse kinematic model

The IKM of Delta PKM consists in finding the corresponding joint positions of the three actuated joints $\mathbf{q} = \begin{bmatrix} q_1 & q_2 & q_3 \end{bmatrix}^T \in \mathbb{R}^{3 \times 1}$ given the position of the traveling-plate, expressed in the fixed reference frame $O - x_o, y_o, z_o$ as $\mathbf{X} = \begin{bmatrix} x & y & z \end{bmatrix}^T \in \mathbb{R}^{3 \times 1}$. As previously mentioned, the IKM of the Delta PKM can be computed based on of the closed-loop equation [94].

$$||{}^o\mathbf{B}_i^o\mathbf{C}_i||^2 = l_i^2, \forall i = 1, 2, 3 \tag{3.23}$$

This equation involves all geometric variables of the Delta PKM. Using Fig. 3.4 the following equation is established in order to obtain the IKM of Delta PKM.

$$ {}^o\mathbf{A}_i = R_b \begin{bmatrix} \cos(\alpha_i) & \sin(\alpha_i) & 0 \end{bmatrix}^T \tag{3.24}$$

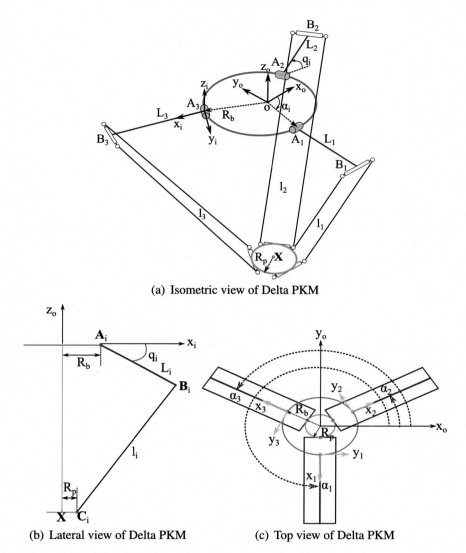

(a) Isometric view of Delta PKM

(b) Lateral view of Delta PKM   (c) Top view of Delta PKM

FIGURE 3.4 Kinematics illustration through different view of Delta PKM.

Where $\mathbf{A}_i$, for $i = 1, 2, 3$ represents the location of the three actuated joints, expressed in the fixed reference frame, R_b is the fixed-base radius. The actuated joints are arranged with the following angles $\boldsymbol{\alpha} = \begin{bmatrix} \frac{3\pi}{2} & \frac{\pi}{6} & \frac{5\pi}{6} \end{bmatrix}^T$.

The points $\mathbf{B}_i$ and $\mathbf{C}_i$ whose coordinates are expressed in the fixed reference frame $O - x_o, y_o, z_o$ are defined as follows:

TABLE 3.1   Summary of the kinematic parameters
of the Delta PKM.

Parameter	Description	Value
L	Rear-arm length	0.3 m
l	Forearm length	0.624 m
R_b	Fixed base radius	0.1267 m
R_p	Traveling plate radius	0.0497 m

$$^o\mathbf{B}_i = {}^o\mathbf{A}_i + L \left[\cos(\alpha_i)\cos(q_i) \quad \sin(\alpha_i)\cos(q_i) \quad -\sin(q_i)\right]^T \qquad (3.25)$$

$$^o\mathbf{C}_i = \left[R_p\cos(\alpha_i) + x \quad R_p\sin(\alpha_i) + y \quad z\right]^T \qquad (3.26)$$

Being L, the rear-arm length, and R_p is the traveling-plate radius. For further analysis, an auxiliary frame located at $^o\mathbf{A}_i$-x_i, y_i, z_i is defined, where the auxiliary vectors $^i\mathbf{x}_i$ and $^i\mathbf{y}_i$ are defined as follows:

$$^i\mathbf{x}_i = \left[\cos(\alpha_i) \quad \sin(\alpha_i) \quad 0\right]^T \qquad (3.27)$$

$$^i\mathbf{y}_i = \left[-\sin(\alpha_i) \quad \cos(\alpha_i) \quad 0\right]^T \qquad (3.28)$$

Considering (3.25), (3.27), and (3.28), we can rewrite expression (3.24) in the following form to obtain the values of the joint variables q_i.

$$D_i\sin(q_i) + E_i\cos(q_i) + F_i = 0 \quad \forall i = 1, 2, 3 \qquad (3.29)$$

where $D_i = 2L_i({}^o\mathbf{A}_i^o\mathbf{C}_i \cdot \mathbf{z}_o)$, $E_i = 2L_i({}^o\mathbf{A}_i^o\mathbf{C}_i \cdot^i \mathbf{x}_i)$, and $F_i = l_i^2 - L_i^2 - ||^o\mathbf{A}_i^o\mathbf{C}_i||^2$. By solving (3.30) for the values of q_i, leads to:

$$q_i = 2\arctan\left(\frac{-D_i \pm \sqrt{\Delta_i}}{F_i - E_i}\right) \qquad (3.30)$$

Being $\Delta_i = D_i^2 + E_i^2 - F_i^2$. The mechanism of our Delta PKM, including all the coupling parts, was designed and assembled in SolidWorks software. The kinematic parameters of our Delta PKM are obtained from its CAD model design. Such parameters are presented in Table 3.1.

3.4.1.1 Application example

It is desired that Delta PKM be positioned in the following coordinates in the workspace whose units are given in meters $\mathbf{X} = \left[-0.1 \quad 0.2 \quad -0.6\right]^T$. Making use of the IKM, we obtain the following position in joint space whose units are provided in radians: $\mathbf{q} = \left[0.813 \quad 0.427 \quad 0.061\right]^T$. Fig. 3.5 represents the obtained pose of the robot.

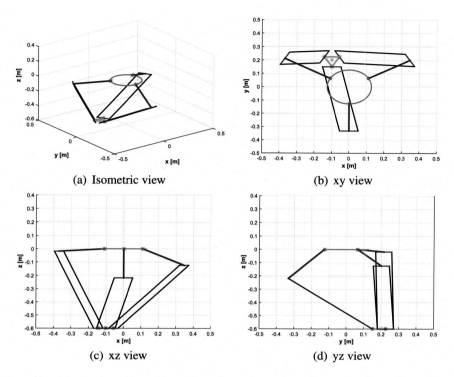

(a) Isometric view

(b) xy view

(c) xz view

(d) yz view

FIGURE 3.5   Illustration of the positioning of Delta PKM by using the IKM.

ALGORITHM 3.1 (Programming code in MATLAB® of the IKM of Delta PKM).

For a better understanding of the equations above, we provide the programming code in MATLAB.

```matlab
function q = IKM_Delta(X)

x=X(1);
y=X(2);
z=X(3);

%%% Kinematic parameters of Delta PKM
L=0.3;
l=0.624;
Rb=0.1267;
Rp=0.0497;
```

```
12  %%% Angles between each kinematic chain
13  alpha1=(3*pi)/2;
14  alpha2=pi/6;
15  alpha3=(5*pi)/6;
16  %%% Actuated joints position
17  A1=Rb*[cos(alpha1);sin(alpha1);0];
18  A2=Rb*[cos(alpha2);sin(alpha2);0];
19  A3=Rb*[cos(alpha3);sin(alpha3);0];
20  %%% Position of the traveling-plate joints
21  C1=[Rp*cos(alpha1)+x;Rp*sin(alpha1)+y;z];
22  C2=[Rp*cos(alpha2)+x;Rp*sin(alpha2)+y;z];
23  C3=[Rp*cos(alpha3)+x;Rp*sin(alpha3)+y;z];
24  % Auxiliary frame x-coordinates
25  X1=[cos(alpha1);sin(alpha1);0];
26  X2=[cos(alpha2);sin(alpha2);0];
27  X3=[cos(alpha3);sin(alpha3);0];
28  z0=[0;0;1];
29  %Computation of auxilary variables
30  D1=2*L*dot(C1-A1,z0);
31  D2=2*L*dot(C2-A2,z0);
32  D3=2*L*dot(C3-A3,z0);
33
34  E1=2*L*dot(C1-A1,X1);
35  E2=2*L*dot(C2-A2,X2);
36  E3=2*L*dot(C3-A3,X3);
37
38  F1=l*l-L*L-norm(C1-A1)^2;
39  F2=l*l-L*L-norm(C2-A2)^2;
40  F3=l*l-L*L-norm(C3-A3)^2;
41
42  Delta1=D1*D1+E1*E1-F1*F1;
43  Delta2=D2*D2+E2*E2-F2*F2;
44  Delta3=D3*D3+E3*E3-F3*F3;
45
46  % Computation of the position of the actuated joints
47  q1=-2*atan((-D1-sqrt(Delta1))/(F1-E1));
48  q2=-2*atan((-D2-sqrt(Delta2))/(F2-E2));
49  q3=-2*atan((-D3-sqrt(Delta3))/(F3-E3));
50
51  q = [q1;q2;q3];
```

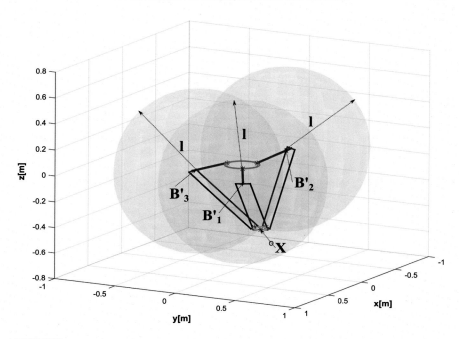

FIGURE 3.6   Illustration of the three virtual spheres intersection used to find $^o\mathbf{X}$ position.

3.4.2 Forward kinematic model

As stated in previous sections of this chapter, it is necessary to establish a set of nonlinear algebraic equations describing the intersection of virtual spheres to find the traveling plate position. Fig. 3.6 illustrates graphically the FKM solution applied Delta PKM. Emphasizing the 3-DOF Delta PKM, its FKM consists in finding the corresponding traveling plate position $^o\mathbf{X} = \begin{bmatrix} x & y & z \end{bmatrix}^T \in \mathbb{R}^{3\times1}$ expressed in the fixed reference frame $O - x_o, y_o, z_o$, given the joint positions of the three actuated joints $\mathbf{q} = \begin{bmatrix} q_1 & q_2 & q_3 \end{bmatrix}^T \in \mathbb{R}^{3\times1}$. Fig. 3.6 illustrates the intersection algorithm of three virtual spheres to obtain the FKM of Delta Robot. A crucial part of applying this method is to establish the virtual sphere centers. Considering the kinematic relationships of a PKM with delta-like architecture, the virtual spheres' centers can be established by following kinematic expression.

$$^o\mathbf{B}'_i =^o \mathbf{B}_i -^X \mathbf{X}_i \tag{3.31}$$

Where $^X\mathbf{X}_i$ is defined in the traveling plate reference frame $X - x_x, y_x, z_x$ as follows:

$$^X\mathbf{X}_i = \begin{bmatrix} R_p \cos(\alpha_i) & R_p \sin(\alpha_i) & 0 \end{bmatrix}^T \tag{3.32}$$

Thus, the sphere equation may be defined as follows:

$$(x - x_i)^2 + (y - y_i)^2 + (z - z_i)^2 = l_i^2, \quad \forall i = 1, 2, 3 \qquad (3.33)$$

Where x_i, y_i, and z_i represents the coordinates of the center of the sphere. The expression (3.32) can be expanded to obtain the following equation:

$$^o\mathbf{B}_i' = \begin{bmatrix} x_i \\ y_i \\ z_i \end{bmatrix} = \begin{bmatrix} (L\cos(q_i) + R_b - R_p)\cos(\alpha_i) \\ (L\cos(q_i) + R_b - R_p)\sin(\alpha_i) \\ -L\sin(q_i) \end{bmatrix} \qquad (3.34)$$

For the 3-DOF Delta PKM, (3.34) is extended to obtain the following system of equations:

$$
\begin{aligned}
x^2 + y^2 + z^2 - 2x_1 x - 2y_1 y - 2z_1 z + w_1 = l^2 \\
x^2 + y^2 + z^2 - 2x_2 x - 2y_2 y - 2z_2 z + w_2 = l^2 \\
x^2 + y^2 + z^2 - 2x_3 x - 2y_3 y - 2z_3 z + w_3 = l^2
\end{aligned} \qquad (3.35)
$$

In which, $w_i = x_i^2 + y_i^2 + z_i^2$ for $i = 1, 2, 3$. The solution of these equations can be expressed as follows:

$$x = d_2 z + e_2 \qquad (3.36)$$
$$y = d_1 z + e_1 \qquad (3.37)$$
$$z = \frac{-b - \sqrt{b^2 - 4ac}}{2a} \qquad (3.38)$$

Where the auxiliary constants used in (3.37)-(3.39) are defined as follows:

$$a_1 = \frac{y_1 - y_2}{x_2 - x_1} \quad b_1 = \frac{z_1 - z_2}{x_2 - x_1} \quad c_1 = \frac{w_2 - w_1}{2(x_2 - x_1)} \quad d_1 = \frac{b_2 - b_1}{a_1 - a_2} \quad e_1 = \frac{c_2 - c_1}{a_1 - a_2}$$

$$a_2 = \frac{y_1 - y_3}{x_3 - x_1} \quad b_2 = \frac{z_1 - z_3}{x_3 - x_1} \quad c_2 = \frac{w_3 - w_1}{2(x_3 - x_1)} \quad d_2 = a_1 d_1 + b_1 \quad e_2 = a_1 e_1 + c_1$$

$$
\begin{aligned}
a &= d_1^2 + d_2^2 + 1 \\
b &= 2[(e_1 - y_1)d_1 + (e_2 - x_1)d_2 - z_1] \\
c &= e_1^2 + e_2^2 - 2(e_1 y_1 + e_2 x_1) + w_1 - l^2
\end{aligned}
$$

3.4.2.1 *Application example*

We want to know the obtained position if all active joints are seated in zero, that is $\mathbf{q} = \begin{bmatrix} 0 & 0 & 0 \end{bmatrix}^T$. Making use of the FKM, we obtain the following position in Cartesian space whose units are provided in meters: $\mathbf{X} = \begin{bmatrix} 0 & 0 & -0.4972 \end{bmatrix}^T$. Fig. 3.7 represents the obtained pose of the robot.

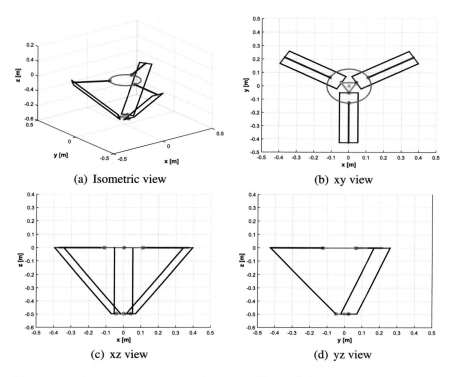

(a) Isometric view   (b) xy view

(c) xz view   (d) yz view

FIGURE 3.7   Illustration of the positioning of Delta PKM using the FKM.

ALGORITHM 3.2 (Programming code in MATLAB of the FKM of Delta PKM).

For a better understanding of the equations above, we provide the programming code in MATLAB.

```matlab
1  function X = FKM_Delta(q)
2
3  q1=q(1);
4  q2=q(2);
5  q3=q(3);
6
7  %%% Kinematic parameters of Delta PKM
8  L=0.3;
9  l=0.624;
10 Rb=0.1267;
11 Rp=0.0497;
```

```matlab
12    %%% Angles between each kinematic chain
13    alpha1=(3*pi)/2;
14    alpha2=pi/6;
15    alpha3=(5*pi)/6;
16
17    %%% Coordinates of the three virtual spheres' centers
18    x1=(L*cos(q1)+Rb-Rp)*cos(alpha1);
19    x2=(L*cos(q2)+Rb-Rp)*cos(alpha2);
20    x3=(L*cos(q3)+Rb-Rp)*cos(alpha3);
21
22    y1=(L*cos(q1)+Rb-Rp)*sin(alpha1);
23    y2=(L*cos(q2)+Rb-Rp)*sin(alpha2);
24    y3=(L*cos(q3)+Rb-Rp)*sin(alpha3);
25
26    z1=-L*sin(q1);
27    z2=-L*sin(q2);
28    z3=-L*sin(q3);
29
30    %%% Squared coordinates
31    w1=x1*x1+y1*y1+z1*z1;
32    w2=x2*x2+y2*y2+z2*z2;
33    w3=x3*x3+y3*y3+z3*z3;
34
35    %%% Auxiliary terms
36    a1=(y1-y2)/(x2-x1);
37    a2=(y1-y3)/(x3-x1);
38    b1=(z1-z2)/(x2-x1);
39    b2=(z1-z3)/(x3-x1);
40    c1=(w2-w1)/(2*(x2-x1));
41    c2=(w3-w1)/(2*(x3-x1));
42    d1=(b2-b1)/(a1-a2);
43    d2=a1*d1+b1;
44    e1=(c2-c1)/(a1-a2);
45    e2=a1*e1+c1;
46
47    a=d1*d1+d2*d2+1;
48    b=2*((e1-y1)*d1+(e2-x1)*d2-z1);
49    c=e1*e1+e2*e2-2*(e1*y1+e2*x1)+w1-l*l;
50
51    %%% FKM solution
52    z=(-b-sqrt(b*b-4*a*c))/(2*a);
53    x=d2*z+e2;
```

```
54  y=d1*z+e1;
55
56
57  X = [x;y;z];
```

3.4.3 Velocity relationship and Jacobian analysis

Jacobian matrices allow us to express the velocities of the actuated joints based-on the velocities of the traveling plate. Having in mind that $||{}^o\mathbf{B}_i^o\mathbf{C}_i||^2 = BC_{ix}^2 + BC_{iy}^2 + BC_{iz}^2$, Eq. (3.14) can be rewritten as follows:

$$x^2 - 2Lx\cos\alpha_i\cos q_i + L^2\cos^2\alpha_i\cos^2 q_i - 2aL\cos\alpha_i\cos q_i + a^2$$
$$+ y^2 - 2Ly\sin\alpha_i\cos q_i + L^2\sin^2\alpha_i\cos^2 q_i - 2bL\sin\alpha_i\cos q_i + b^2 \qquad (3.39)$$
$$+ z^2 + 2Lz\sin q_i + L^2\sin^2 q_i = l$$

Where $a = R_p\cos\alpha_i - R_b\cos\alpha_i$ and $b = R_p\sin\alpha_i - R_b\sin\alpha_i$. The time derivative of (3.40) yields the general equation of velocities' relationship expressed in the following form:

$$((R_p - R_b)\cos\alpha_i + x - L\cos\alpha_i\cos q_i)\dot{x}$$
$$+ ((R_p - R_b)\sin\alpha_i + y - L\sin\alpha_i\cos q_i)\dot{y} + (z + L\sin q_i)\dot{z} \qquad (3.40)$$
$$= -L((x\cos\alpha_i + y\sin\alpha_i + R_p - R_b)\sin q_i + z\cos q_i)\dot{q}_i \quad \forall i = 1,...,n$$

The terms of the previous expression are arranged with respect to the velocities in Cartesian and joint space in order to obtain the Jacobian matrices $\mathbf{J}_x \in \mathbb{R}^{3\times 3}$, and $\mathbf{J}_q \in \mathbb{R}^{3\times 3}$. Therefore, the general Jacobian matrix of the 3-DOF Delta robot may be written as:

$$\mathbf{J} = \mathbf{J}_q^{-1}\mathbf{J}_x \in \mathbb{R}^{3\times 3} \qquad (3.41)$$

Considering (3.41), the Jacobian matrices for the Delta PKM can be expressed as follows:

$$\mathbf{J}_x = \begin{bmatrix} R\cos\alpha_1 + x - L\cos\alpha_1\cos q_1 & R\sin\alpha_1 + y - L\sin\alpha_1\cos q_1 & z + L\sin q_1 \\ R\cos\alpha_2 + x - L\cos\alpha_2\cos q_2 & R\sin\alpha_2 + y - L\sin\alpha_2\cos q_2 & z + L\sin q_2 \\ R\cos\alpha_3 + x - L\cos\alpha_3\cos q_3 & R\sin\alpha_3 + y - L\sin\alpha_3\cos q_3 & z + L\sin q_3 \end{bmatrix}$$
$$(3.42)$$

$$\mathbf{J}_q = \begin{bmatrix} Jq_{11} & 0 & 0 \\ 0 & Jq_{22} & 0 \\ 0 & 0 & Jq_{33} \end{bmatrix} \qquad (3.43)$$

where $R = R_p - R_b$ and

$$Jq_{11} = -L((x \cos \alpha_1 + y \sin \alpha_1 + R) \sin q_1 + z \cos q_1)$$
$$Jq_{22} = -L((x \cos \alpha_2 + y \sin \alpha_2 + R) \sin q_2 + z \cos q_2)$$
$$Jq_{33} = -L((x \cos \alpha_3 + y \sin \alpha_3 + R) \sin q_3 + z \cos q_3)$$

The joint velocities of the Delta PKM $\dot{q} \in \mathbb{R}^{3 \times 1}$ may be computed from the Cartesian velocities $\dot{X} \in \mathbb{R}^{3 \times 1}$ through the general Jacobian matrix $J \in \mathbb{R}^{3 \times 3}$ as described in previous subsections as follows:

$$\dot{q} = J\dot{X} \tag{3.44}$$

To obtain the Cartesian velocities for the 3-DOF Delta robot in function of the joint velocities (3.44) is reformulated as:

$$\dot{X} = J_m \dot{q} \tag{3.45}$$

being J_m the inverse matrix of J.

ALGORITHM 3.3 (Programming code in MATLAB of the Jacobian matrix of Delta PKM).
     For a better understanding of the equations above, we provide the programming code in MATLAB.

```
1   function J = jac_Delta(q,X)
2   x=X(1);
3   y=X(2);
4   z=X(3);
5
6   q1=Q(1);
7   q2=Q(2);
8   q3=Q(3);
9   %%% Kinematic parameters of Delta PKM
10  L=0.3;
11  l=0.624;
12  Rb=0.1267;
13  Rp=0.0497;
14  R=Rp-Rb;
15  %%% Angles between each kinematic chain
16  alpha1=(3*pi)/2;
17  alpha2=pi/6;
18  alpha3=(5*pi)/6;
19
```

```
20  jx11=R*cos(alpha1)+x−L*cos(alpha1)*cos(q1);
21  jx12=R*sin(alpha1)+y−L*sin(alpha1)*cos(q1);
22  jx13=z+L*sin(q1);
23
24  jx21=R*cos(alpha2)+x−L*cos(alpha2)*cos(q2);
25  jx22=R*sin(alpha2)+y−L*sin(alpha2)*cos(q2);
26  jx23=z+L*sin(q2);
27
28  jx31=R*cos(alpha3)+x−L*cos(alpha3)*cos(q3);
29  jx32=R*sin(alpha3)+y−L*sin(alpha3)*cos(q3);
30  jx33=z+L*sin(q3);
31
32  Jx=[jx11 jx12 jx13;jx21 jx22 jx23;jx31 jx32 jx33];
33
34  jq11=−L*(sin(q1)*(x*cos(alpha1)+y*sin(alpha1)+R)+z*cos(q1
        ));
35  jq22=−L*(sin(q2)*(x*cos(alpha2)+y*sin(alpha2)+R)+z*cos(q2
        ));
36  jq33=−L*(sin(q3)*(x*cos(alpha3)+y*sin(alpha3)+R)+z*cos(q3
        ));
37
38  Jq=[jq11 0 0;0 jq22 0;0 0 jq33];
39
40  J = inv(Jq)*Jx;
```

3.4.4 Inverse dynamic model

The simplified IDM of the 3-DOF Delta PKM is obtained considering the modeling simplification described previously, where the forearms' mass is distributed equivalently between the rear-arms and the traveling plate. Moreover, due to the complexity to precisely compute the friction effects, they are not taken into account in this dynamic model. Therefore, Eq. (3.12) can be rewritten as:

$$\mathbf{\Gamma}(t) = \mathbf{\Gamma}_{ra}(t) + \mathbf{\Gamma}_{fa}(t) + \mathbf{\Gamma}_{tp}(t) \qquad (3.46)$$

Where $\mathbf{\Gamma}_{ra}(t)$ represents the vector of torques produced by the rear-arms, $\mathbf{\Gamma}_{fa}(t)$ are the toques generated by the forearms, and $\mathbf{\Gamma}_{tp}(t)$ are the torques produced by the traveling plate. The rear-arms torques are computed through the following equation:

$$\mathbf{\Gamma}_{ra}(t) = \mathbf{I}_{act}\ddot{\mathbf{q}} + \mathbf{I}_{ra}\ddot{\mathbf{q}} - \mathbf{M}_{ra}gL_c\cos(\mathbf{q}) \qquad (3.47)$$

Where $\mathbf{I}_{act} = \mathrm{diag}([I_{act}]) \in \mathbb{R}^{3\times3}$ is a diagonal matrix containing the actuators including the gear-head, $\mathbf{I}_{ra} = \mathrm{diag}([I_{ra}]) \in \mathbb{R}^{3\times3}$ is the inertia matrix of the rear-arms, $\cos(\mathbf{q}) \in \mathbb{R}^{3\times1}$ is a vector containing the cosine of each angle q_i, $i = 1...3$, $\mathbf{M}_{ra} = \mathrm{diag}([m_{ra}]) \in \mathbb{R}^{3\times3}$ is the mass matrix of the rear-arms, and L_c is the distance from the rotational axis of the rear-arm to its gravity center. The contributions of the inertia and torque due to the spatial parallelograms are given by means of the following expression:

$$\mathbf{\Gamma}_{fa}(t) = \mathbf{I}_{fa}\ddot{\mathbf{q}} - \mathbf{M}_{fa}gL\cos(\mathbf{q}) + \mathbf{J}_m^T\mathbf{M}_{nfa}(\ddot{\mathbf{X}} + \mathbf{g}) \qquad (3.48)$$

Being $\mathbf{I}_{fa} = \mathrm{diag}([L^2\frac{m_{fa}}{2}]) \in \mathbb{R}^{3\times3}$, and $\mathbf{M}_{fa} = \mathrm{diag}([\frac{m_{fa}}{2}]) \in \mathbb{R}^{3\times3}$, where m_{fa} is the mass of a forearm considering the two parallel bars.

$\mathbf{M}_{nfa} = \mathrm{diag}([3\frac{m_{fa}}{2}]) \in \mathbb{R}^{3\times3}$ is a diagonal matrix containing the half masses of the forearms multiplied by $n = 3$, $\mathbf{J}_m \in \mathbb{R}^{3\times3}$ is the inverse Jacobian matrix, and $\mathbf{g} = \begin{bmatrix} 0 & 0 & g \end{bmatrix} \in \mathbb{R}^{3\times1}$ is the vector of the gravity acceleration. The contributions of the inertial and gravity forces acting on the traveling plate to the actuators can be computed by means of the following expression:

$$\mathbf{\Gamma}_{tp}(t) = \mathbf{J}_m^T\mathbf{M}_{tp}(\ddot{\mathbf{X}} + \mathbf{g}) \qquad (3.49)$$

Being $\mathbf{M}_{tp} = \mathrm{diag}([m_{tp}]) \in \mathbb{R}^{3\times3}$, where m_{tp} is the traveling plate's mass. Substituting Eqs. (3.48)–(3.50) into (3.47), we obtain the following expression of the Delta PKM dynamic model.

$$\mathbf{\Gamma} = (\mathbf{I}_{act} + \mathbf{I}_{ra} + \mathbf{I}_{fa})\ddot{\mathbf{q}} - (\mathbf{M}_{ra}gL_G + \mathbf{M}_{fa}gL)\cos(\mathbf{q}) + \mathbf{J}_m^T(\mathbf{M}_{tp} + \mathbf{M}_{nfa})(\ddot{\mathbf{X}} - \mathbf{G}) \qquad (3.50)$$

To express this IDM entirely in terms of joint variables we must derive (3.46), obtaining:

$$\ddot{\mathbf{X}} = \mathbf{J}_m\ddot{\mathbf{q}} + \dot{\mathbf{J}}_m\dot{\mathbf{q}} \qquad (3.51)$$

Eq. (3.52) is substituted into (3.51); the resulted expression is rearranged in the following form:

$$\mathbf{M}(\mathbf{q})\ddot{\mathbf{q}} + \mathbf{C}(\mathbf{q}, \dot{\mathbf{q}})\dot{\mathbf{q}} + \mathbf{G}(\mathbf{q}) = \mathbf{\Gamma} \qquad (3.52)$$

The inertia matrix $\mathbf{M}(\mathbf{q}) \in \mathbb{R}^{3\times3}$ is expressed as follows:

$$\mathbf{M}(\mathbf{q}) = \mathbf{I}_t + \mathbf{J}_m^T\mathbf{M}_t\mathbf{J}_m \qquad (3.53)$$

where $\mathbf{I}_t = \mathbf{I}_{act} + \mathbf{I}_{ra} + \mathbf{I}_{fa} \in \mathbb{R}^{3\times3}$ and $\mathbf{M}_t = \mathbf{M}_{tp} + \mathbf{M}_{nfa} \in \mathbb{R}^{3\times3}$. The matrix of Coriolis and centrifugal forces $\mathbf{C}(\mathbf{q}, \dot{\mathbf{q}}) \in \mathbb{R}^{3\times3}$ is expressed by:

$$\mathbf{C}(\mathbf{q}, \dot{\mathbf{q}}) = \mathbf{J}_m^T\mathbf{M}_t\dot{\mathbf{J}}_m \qquad (3.54)$$

TABLE 3.2 Summary of the kinematic parameters of the Delta PKM.

Parameter	Description	Value
m_{tp}	Traveling plate mass	0.19 kg
m_{ra}	Rear-arm mass	0.29 kg
m_{fa}	Forearm mass	0.28 kg
I_{ra}	Rear-arm inertia	0.0213 kg m^2
I_{act}	Motor inertia	3.8×10^{-6} kg m^2

Finally, the gravity vector $\mathbf{G}(\mathbf{q}) \in \mathbb{R}^{3 \times 1}$ is given by:

$$\mathbf{G}(\mathbf{q}) = \mathbf{J}_m^T \mathbf{M}_t \mathbf{g} + (\mathbf{M}_{ra} g L_c + \mathbf{M}_{fa} g L) \cos(\mathbf{q}) \qquad (3.55)$$

The dynamic parameters of Delta PKM as the masses of the rear-arms, forearms, the traveling plate, coupling parts, and rear-arms inertia were calculated using the material assignation functionality of SolidWorks software. The other dynamic parameters as the inertias and masses of the motors were obtained from the datasheets of their manufacturers. The values of such dynamical parameters are provided in Table 3.2.

3.5 Application to modeling algorithms to 5-DOF SPIDER4 RA-PKM

SPIDER4 [109], [48] (cf. Fig. 3.1) is a novel RA-PKM designed and manufactured within a collaboration between The Laboratoire d'Informatique, de Robotique et de Microélectronique de Montpellier (LIRMM), and Tecnalia company (www.tecnalia.com). This machine is designed for CNC machining tasks of resin materials. It has 5-DOFs: 3 translations denoted by (x, y, z) and 2 rotations denoted by (ϕ, ψ). The translational motion is performed by the parallel structure consisting of four kinematic chains, each one formed by one arm and one forearm, linked by a passive universal joint. Each arm is mounted directly on an electric motor located at the fixed base, and the forearm is connected to the moving platform through two parallel bars, in a Delta-like PKM architecture. The rotational motions are generated by an independent serial wrist mechanism placed on the traveling plate, and actuated by two electric motors. The machining spindle is placed at the end of this mechanism. Fig. 3.8 shows SPIDER4 RA-PKM and its kinematic configuration diagram, where the gray boxes represent the active rotational joints, and the white boxes represent the passive universal joints. To perform a machining task, SPIDER4 makes use of a tooling plate to fix the raw material. The maximum dimensions for the material to be machined should be within a cylinder of 700 mm

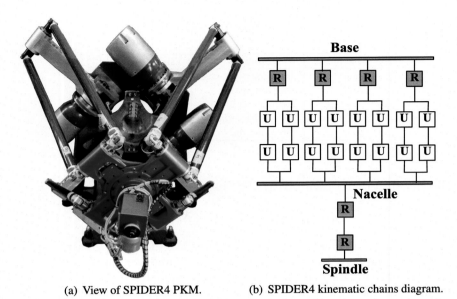

(a) View of SPIDER4 PKM.   (b) SPIDER4 kinematic chains diagram.

FIGURE 3.8   Illustration of SPIDER4 RA-PKM and its kinematics.

diameter and 300 mm depth. The machine and its tooling plate are inside an enclosure for operator safety; its dimensions are 3.9x2.5x2.4 m (LxWxH), as illustrated in Fig. 3.9. It is worth mentioning that SPIDER4 can be equipped with linear encoders to directly measure the traveling plate position from the origin of the fixed base. However, for the control schemes developed in this research, such sensors will not be considered; only the rotary encoders integrated in the motors of the robot will be used. The different components of SPIDER4 are described in detail in Fig. 3.10.

3.5.1 Inverse kinematic model

Let us start with the development of the IKM of SPIDER4, which consists in finding the generalized coordinates vector $\mathbf{Q} = \begin{bmatrix} q_1 & q_2 & q_3 & q_4 & \phi & \psi \end{bmatrix}^T \in \mathbb{R}^{6\times1}$ given the spindle position in the fixed reference frame $O - x_o, y_o, z_o$ as $^o\mathbf{S}_S = \begin{bmatrix} x & y & z & \phi & \psi \end{bmatrix}^T \in \mathbb{R}^{5\times1}$. It is worth mentioning that the variables ϕ, ψ are the same for operational and joint spaces; in addition, it is possible to define a vector which involves only the parallel structure joint variables $\mathbf{q} = \begin{bmatrix} q_1 & q_2 & q_3 & q_4 \end{bmatrix}^T \in \mathbb{R}^{4\times1}$. SPIDER4 RA-PKM shares several similarities with the Delta PKM. Indeed, the kinematic composition of both PKMs is composed of one rear-arm and one forearm formed by two parallel bars. Nevertheless, SPIDER4 has a wrist mechanism on the traveling plate. Therefore, before to formulate

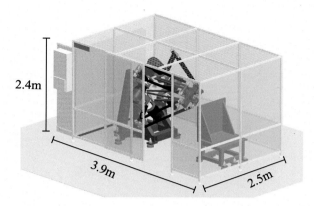

FIGURE 3.9   Illustration of the enclosure dimensions of the working area of SPIDER4 RA-PKM.

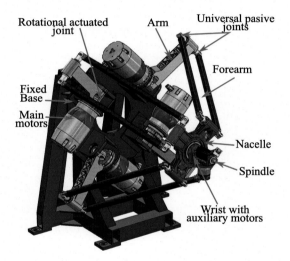

FIGURE 3.10   CAD-view of SPIDER4 RA-PKM and its main components.

the kinematic equations of the delta-like positioning device, it is essential to develop the kinematic relations of the wrist to get the position of the traveling plate denoted by $^{o}\mathbf{N}_N = \begin{bmatrix} x_n & y_n & z_n \end{bmatrix}^{T} \in \mathbb{R}^{3 \times 1}$. Considering the kinematic diagram of SPIDER4 depicted in Fig. 3.11, we can establish the following rotation matrices to derive the kinematics relationships of the wrist mechanism.

$$\mathbf{R}_Z = \begin{bmatrix} \cos(\phi) & -\sin(\phi) & 0 \\ \sin(\phi) & \cos(\phi) & 0 \\ 0 & 0 & 1 \end{bmatrix}, \ \mathbf{R}_Y = \begin{bmatrix} \cos(\psi) & 0 & \sin(\psi) \\ 0 & 1 & 0 \\ -\sin(\psi) & 0 & \cos(\psi) \end{bmatrix} \tag{3.56}$$

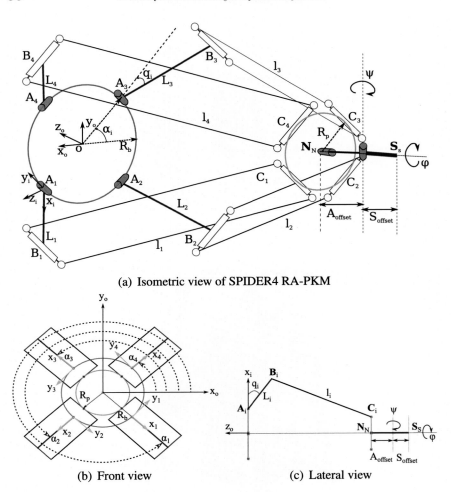

(a) Isometric view of SPIDER4 RA-PKM

(b) Front view   (c) Lateral view

FIGURE 3.11   Kinematics illustration of SPIDER4 RA-PKM.

The above expressions involve the wrist mechanism variables (ϕ, ψ), consequently $^o\mathbf{N}_N$ can be defined as follows:

$$^o\mathbf{N}_N = \begin{bmatrix} x \\ y \\ z + A_{offset} \end{bmatrix} + \mathbf{R}_Z \mathbf{R}_Y \begin{bmatrix} 0 \\ 0 \\ S_{offset} \end{bmatrix} \qquad (3.57)$$

By rewriting Eq. (3.58); $^o\mathbf{N}_N$ can be expressed as:

$$^o\mathbf{N}_N = \begin{bmatrix} x + S_{offset}\cos(\phi)\sin(\psi) \\ y + S_{offset}\sin(\phi)\sin(\psi) \\ z + S_{offset}\cos(\psi) + A_{offset} \end{bmatrix} \tag{3.58}$$

Where S_{offset} is the distance between the ψ joint axis and the tip of the spindle (Fig. 3.11), and A_{offset} is the distance between the center of the traveling plate and the ψ joint axis (Fig. 3.11). Once $^o\mathbf{N}_N$ is computed, we can set the active and passive joint locations with respect to the Cartesian-fixed frame $O - x_o, y_o, z_o$. The actuated joints located at points $\mathbf{A}_i$, $i = 1...4$, are expressed with respect to the fixed reference frame by the following vector:

$$^o\mathbf{A}_i = R_b \begin{bmatrix} \cos(\alpha_i) & \sin(\alpha_i) & 0 \end{bmatrix}^T \tag{3.59}$$

Where R_b is the fixed-base radius. The four main actuators are placed on the fixed base with the following angles with respect to the fixed Cartesian frame $\boldsymbol{\alpha} = \begin{bmatrix} \frac{7\pi}{4} & \frac{5\pi}{4} & \frac{3\pi}{4} & \frac{\pi}{4} \end{bmatrix}^T$. The universal passive joints are located at points $\mathbf{B}_i$ and $\mathbf{C}_i$ whose coordinates are expressed in the fixed reference frame $O - x_o, y_o, z_o$ as follows:

$$^o\mathbf{B}_i = {}^o\mathbf{A}_i + L \begin{bmatrix} \cos(\alpha_i)\cos(q_i) & \sin(\alpha_i)\cos(q_i) & -\sin(q_i) \end{bmatrix}^T \tag{3.60}$$

$$^o\mathbf{C}_i = \begin{bmatrix} R_p\cos(\alpha_i) + x_n & R_p\sin(\alpha_i) + y_n & z_n \end{bmatrix}^T \tag{3.61}$$

being L the arm length and R_p the traveling plate radius. For further analysis, an auxiliary frame located at $^o\mathbf{A}_i$-x_i, y_i, z_i is defined, where the two following auxiliary vectors $^i\mathbf{x}_i$ and $^i\mathbf{y}_i$ are defined:

$$^i\mathbf{x}_i = \begin{bmatrix} \cos(\alpha_i) & \sin(\alpha_i) & 0 \end{bmatrix}^T \tag{3.62}$$

$$^i\mathbf{y}_i = \begin{bmatrix} -\sin(\alpha_i) & \cos(\alpha_i) & 0 \end{bmatrix}^T \tag{3.63}$$

The expression relating the operational traveling plate variables $^o\mathbf{N}_N$ to the parallel structure joint variables $\mathbf{q}$ is called the close-loop equation, which can be expressed as:

$$||^o\mathbf{C}_i - {}^o\mathbf{B}_i||^2 = l_i^2 \tag{3.64}$$

According to [95], [38] the mathematical expression (3.9) can be rewritten as follows in order to obtain the values of the joint variables $\mathbf{q}$:

$$D_i \sin(q_i) + E_i \cos(q_i) + F_i = 0 \quad \forall i = 1, 2, 3, 4 \tag{3.65}$$

where $D_i = 2L_i({}^o\mathbf{A}_i^o\mathbf{C}_i \cdot \mathbf{z}_o)$, $E_i = 2L_i({}^o\mathbf{A}_i^o\mathbf{C}_i \cdot^i \mathbf{x}_i)$, and $F_i = l_i^2 - L_i^2 - ||^o\mathbf{A}_i^o\mathbf{C}_i||^2$. The values of q_i are then obtained by solving (3.66), resulting in the following expression:

$$q_i = 2\arctan\left(\frac{-D_i \pm \sqrt{\Delta_i}}{F_i - E_i}\right) \tag{3.66}$$

TABLE 3.3 Summary of the kinematic parameters of SPIDER4.

Parameter	Description	Value
L	Rear-arm length	0.535 m
l	Forearm length	1.100 m
R_b	Fixed base radius	0.4 m
R_p	traveling plate radius	0.260 m
S_{offset}	Distance between ψ and ^oS_S	1.135 m
A_{offset}	Distance between ^oN_N and ψ	0.198375 m

Being $\Delta_i = D_i^2 + E_i^2 - F_i^2$. The kinematic parameters of SPIDER4 as the lengths of the limbs, the radius of the fixed base and the traveling plate were obtained from the CAD model provided by Tecnalia Company. Thereby, the kinematic parameters of SPIDER4 are known with enough precision. These parameters are presented in Table 3.3.

3.5.1.1 Application example

It is desired that spindle of SPIDER4 RA-PKM be positioned in the following position $^oS_S = \begin{bmatrix} 0 & 0 & -1.65 & 0 & 0 \end{bmatrix}^T$. By using the IKM we obtain the following position in Cartesian space for the traveling plate $^oN_N = \begin{bmatrix} 0 & 0 & -1.11 \end{bmatrix}^T$. Therefore, the resulting joint space positions are $Q = \begin{bmatrix} 0.4133 & 0.4133 & 0.4133 & 0.4133 & 0 & 0 \end{bmatrix}^T$. Fig. 3.12 represents the obtained pose of SPIDER4.

ALGORITHM 3.4 (Programming code in MATLAB of the IKM of SPIDER4 RA-PKM).

For a better understanding of the equations above, we provide the programming code in MATLAB.

```
1   function [Q, q] = IKM_Spider4(S)
2
3   x=S(1);
4   y=S(2);
5   z=S(3);
6   phi=S(4);
7   psi=S(5);
8
9   %%% Kinematic parameters of SPIDER4 RA-PKM
10  L=0.535;
11  l=1.10;
```

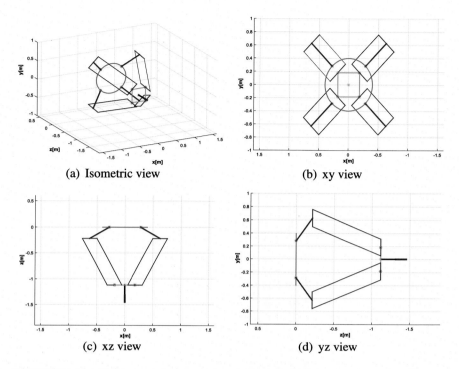

(a) Isometric view

(b) xy view

(c) xz view

(d) yz view

FIGURE 3.12 Illustration of the positioning of SPIDER4 RA-PKM by using the IKM.

```
12   Rb=0.4;
13   Rp=0.26;
14   Aof=0.198375;
15   Sof=0.135;
16
17   %%% Angles between each kinematic chain
18   alpha1=7*pi/4;
19   alpha2=5*pi/4;
20   alpha3=3*pi/4;
21   alpha4=pi/4;
22
23   % Rotation matrices
24
25   Rz=[cos(phi) -sin(phi) 0;sin(phi) cos(phi) 0;0 0 1];
26   Ry=[cos(psi) 0 sin(psi);0 1 0;-sin(psi) 0 cos(psi)];
27
```

```matlab
28   % Computation of the position in cartesian space of the
         traveling plate
29
30   N=[x;y;z+Aof]+Rz*Ry*[0;0;Sof];
31
32   xn=N(1);
33   yn=N(2);
34   zn=N(3);
35
36   %%% Actuated joints position
37   A1=Rb*[cos(alpha1);sin(alpha1);0];
38   A2=Rb*[cos(alpha2);sin(alpha2);0];
39   A3=Rb*[cos(alpha3);sin(alpha3);0];
40   A4=Rb*[cos(alpha4);sin(alpha4);0];
41   %%% Position of the traveling-plate joints
42   C1=[Rp*cos(alpha1)+xn;Rp*sin(alpha1)+yn;zn];
43   C2=[Rp*cos(alpha2)+xn;Rp*sin(alpha2)+yn;zn];
44   C3=[Rp*cos(alpha3)+xn;Rp*sin(alpha3)+yn;zn];
45   C4=[Rp*cos(alpha4)+xn;Rp*sin(alpha4)+yn;zn];
46   % Auxiliary frame x-coordinates
47   X1=[cos(alpha1);sin(alpha1);0];
48   X2=[cos(alpha2);sin(alpha2);0];
49   X3=[cos(alpha3);sin(alpha3);0];
50   X4=[cos(alpha4);sin(alpha4);0];
51   z0 =[0;0;1];
52   %Computation of auxilary variables
53   D1=2*L*dot(C1-A1,z0);
54   D2=2*L*dot(C2-A2,z0);
55   D3=2*L*dot(C3-A3,z0);
56   D4=2*L*dot(C4-A4,z0);
57
58   E1=2*L*dot(C1-A1,X1);
59   E2=2*L*dot(C2-A2,X2);
60   E3=2*L*dot(C3-A3,X3);
61   E4=2*L*dot(C4-A4,X4);
62
63   F1=l*l-L*L-norm(C1-A1)^2;
64   F2=l*l-L*L-norm(C2-A2)^2;
65   F3=l*l-L*L-norm(C3-A3)^2;
66   F4=l*l-L*L-norm(C4-A4)^2;
67
68   Delta1=D1*D1+E1*E1-F1*F1;
```

```
69    Delta2=D2*D2+E2*E2−F2*F2;
70    Delta3=D3*D3+E3*E3−F3*F3;
71    Delta4=D4*D4+E4*E4−F4*F4;
72
73    % Computation of the position of the actuated joints
74    q1=−2*atan((−D1−sqrt(Delta1))/(F1−E1));
75    q2=−2*atan((−D2−sqrt(Delta2))/(F2−E2));
76    q3=−2*atan((−D3−sqrt(Delta3))/(F3−E3));
77    q4=−2*atan((−D4−sqrt(Delta4))/(F4−E4));
78
79    q = [q1;q2;q3;q4];
80
81    Q=[q;phi;psi];
```

3.5.2 Forward kinematic model

For SPIDER4, the FKM is computed in two steps; the first one is to find $^{o}\mathbf{N}_N \in \mathbb{R}^{3 \times 1}$ when $\mathbf{q} \in \mathbb{R}^{4 \times 1}$ is given; in other words, we have to compute the FKM of the delta-like positioning device. The second step lies in computing the spindle position $^{o}\mathbf{S}_S$ once $^{o}\mathbf{N}_N$ has been found. Fig. 3.13 illustrates the intersection algorithm of four virtual spheres to obtain the FKM of SPIDER4. According to this figure, one can see that this method leads to two solutions; however, it is easy to choose the appropriate one for the FKM considering the geometry of the robot. The centers of the virtual spheres for SPIDER4 are established by means of the following kinematic expression:

$$^{o}\mathbf{B}'_i = {}^{o}\mathbf{B}_i - {}^{N}\mathbf{N}_i \quad \forall i = 1, ..., 4 \tag{3.67}$$

Where:

$$^{N}\mathbf{N}_i = \begin{bmatrix} R_p \cos(\alpha_i) & R_p \sin(\alpha_i) & 0 \end{bmatrix}^T. \tag{3.68}$$

The equations of the four virtual spheres needed to obtain the FKM of SPIDER4 delta-like positioning device are given as follows:

$$\begin{aligned}
x_n^2 + y_n^2 + z_n^2 - 2x_1 x_n - 2y_1 y_n - 2z_1 z_n + w_1 &= l^2 \\
x_n^2 + y_n^2 + z_n^2 - 2x_2 x_n - 2y_2 y_n - 2z_2 z_n + w_2 &= l^2 \\
x_n^2 + y_n^2 + z_n^2 - 2x_3 x_n - 2y_3 y_n - 2z_3 z_n + w_3 &= l^2 \\
x_n^2 + y_n^2 + z_n^2 - 2x_4 x_n - 2y_4 y_n - 2z_4 z_n + w_4 &= l^2
\end{aligned} \tag{3.69}$$

In which, x_i, y_i, z_i, for $i = 1, ..., 4$, represent the coordinates of the center of each virtual sphere taken from (3.68), and $w_i = x_i^2 + y_i^2 + z_i^2$ for $i = 1, ..., 4$.

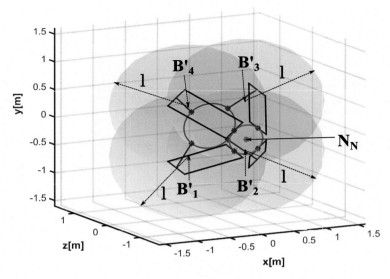

FIGURE 3.13   Illustration of the four virtual spheres intersection used to find ^oN_N position for SPIDER4.

The solution of the system of equations (3.70) corresponding to the traveling plate position is given as follows:

$$x_n = d_2 z_n + e_2 \tag{3.70}$$

$$y_n = d_1 z_n + e_1 \tag{3.71}$$

$$z_n = \frac{-b - \sqrt{b^2 - 4ac}}{2a} \tag{3.72}$$

Where the auxiliary constants involved in expressions (3.71)-(3.73) are defined as follows:

$$a_1 = \frac{y_1 - y_3}{x_3 - x_1} \quad b_1 = \frac{z_1 - z_3}{x_3 - x_1} \quad c_1 = \frac{w_3 - w_1}{2(x_3 - x_1)} \quad d_1 = \frac{b_2 - b_1}{a_1 - a_2} \quad e_1 = \frac{c_2 - c_1}{a_1 - a_2}$$

$$a_2 = \frac{y_2 - y_3}{x_3 - x_1} \quad b_2 = \frac{z_2 - z_3}{x_3 - x_2} \quad c_2 = \frac{w_3 - w_2}{2(x_3 - x_2)} \quad d_2 = a_1 d_1 + b_1 \quad e_2 = a_1 e_1 + c_1$$

$$a = d_1^2 + d_2^2 + 1$$
$$b = 2[(e_1 - y_4)d_1 + (e_2 - x_4)d_2 - z_4]$$
$$c = e_1^2 + e_2^2 - 2(e_1 y_4 + e_2 x_4) + w_4 - l^2$$

Once the positions of the traveling plate (x_n, y_n, z_n) are found, an auxiliary vector is defined in order to find the spindle position. This auxiliary vector involves the variables of the wrist mechanism (ϕ, ψ), and is expressed by:

$$^{o}\mathbf{N}_{aux} = {}^{o}\mathbf{N}_{N} - \begin{bmatrix} S_{offset}\cos(\phi)\sin(\psi) \\ S_{offset}\sin(\phi)\sin(\psi) \\ S_{offset}\cos(\psi) + A_{offset} \end{bmatrix} \quad (3.73)$$

Finally, the spindle position can be defined through the following expression:

$$^{o}\mathbf{S}_{S} = \begin{bmatrix} {}^{o}\mathbf{N}_{aux}^{T} & \phi & \psi \end{bmatrix}^{T} \quad (3.74)$$

3.5.2.1 Application example

Let us compute the resulting positioning of the traveling plate of SPIDER4 given the following vector of joint space positions $\mathbf{Q} = \begin{bmatrix} 0.53 & 1.16 & 1.35 & 0.85 & 0 & 1.57 \end{bmatrix}^{T}$. By using the FKM we obtain the following position in Cartesian space for the traveling plate $^{o}\mathbf{N}_{N} = \begin{bmatrix} 0.43 & -0.2 & -1.35 \end{bmatrix}^{T}$, and $^{o}\mathbf{S}_{S} = \begin{bmatrix} 0.1 & -0.2 & -1.55 & 0 & 1.57 \end{bmatrix}^{T}$ for the position of the spindle tool. Fig. 3.14 represents the obtained pose of SPIDER4.

ALGORITHM 3.5 (Programming code in MATLAB of the FKM of SPIDER4 RA-PKM).

For a better understanding of the equations above, we provide the programming code in MATLAB.

```
1   function [S,N] = FKM_Spider4(Q)
2
3   q1=Q(1);
4   q2=Q(2);
5   q3=Q(3);
6   q4=Q(4);
7   phi=Q(5);
8   psi=Q(6);
9
10  %Kinematic parameters of SPIDER4
11  L=0.535;
12  l=1.10;
13  Rb=0.4;
14  Rp=0.26;
15  Aof=0.198375;
16  Sof=0.135;
17
```

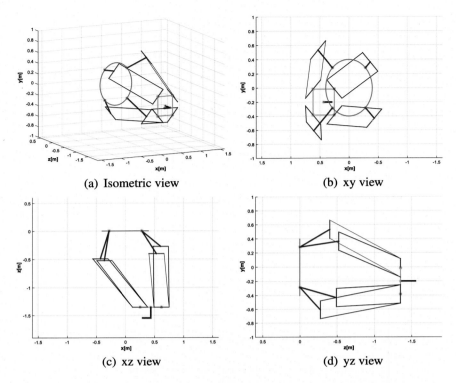

(a) Isometric view   (b) xy view

(c) xz view   (d) yz view

FIGURE 3.14   Illustration of the positioning of SPIDER4 RA-PKM by using the FKM.

```
18  %Angles between each kinematic chain
19  alpha1=7*pi/4;
20  alpha2=5*pi/4;
21  alpha3=3*pi/4;
22  alpha4=pi/4;
23
24  % Coordinates of the four virtual spheres' centers
25  x1=Rb*cos(alpha1)+L*cos(alpha1)*cos(q1)-Rp*cos(alpha1);
26  x2=Rb*cos(alpha2)+L*cos(alpha2)*cos(q2)-Rp*cos(alpha2);
27  x3=Rb*cos(alpha3)+L*cos(alpha3)*cos(q3)-Rp*cos(alpha3);
28  x4=Rb*cos(alpha4)+L*cos(alpha4)*cos(q4)-Rp*cos(alpha4);
29
30  y1=Rb*sin(alpha1)+L*sin(alpha1)*cos(q1)-Rp*sin(alpha1);
31  y2=Rb*sin(alpha2)+L*sin(alpha2)*cos(q2)-Rp*sin(alpha2);
32  y3=Rb*sin(alpha3)+L*sin(alpha3)*cos(q3)-Rp*sin(alpha3);
33  y4=Rb*sin(alpha4)+L*sin(alpha4)*cos(q4)-Rp*sin(alpha4);
```

```
34
35   z1=−L∗sin(q1);
36   z2=−L∗sin(q2);
37   z3=−L∗sin(q3);
38   z4=−L∗sin(q4);
39
40   % Squared Coordinates
41   w1=x1^2+y1^2+z1^2;
42   w2=x2^2+y2^2+z2^2;
43   w3=x3^2+y3^2+z3^2;
44   w4=x4^2+y4^2+z4^2;
45
46   % Auxiliary terms
47   a1=(y1−y3)/(x3−x1);
48   a2=(y2−y3)/(x3−x2);
49   b1=(z1−z3)/(x3−x1);
50   b2=(z2−z3)/(x3−x2);
51   c1=(w3−w1)/(2∗(x3−x1));
52   c2=(w3−w2)/(2∗(x3−x2));
53
54   d1=(b2−b1)/(a1−a2);
55   d2=(a1∗d1+b1);
56   e1=(c2−c1)/(a1−a2);
57   e2=(a1∗e1+c1);
58
59   a=d1^2+d2^2+1;
60   b=2∗(d1∗e1+d2∗e2−x4∗d2−y4∗d1−z4);
61   c=e1^2+e2^2−2∗x4∗e2−2∗y4∗e1−l^2+w4;
62   z=(−b−sqrt(b∗b−4∗a∗c))/(2∗a);
63
64   y=d1∗z+e1;
65   x=d2∗z+e2;
66
67   % Position of the traveling plate
68   N = [x;y;z];
69
70   %Wrist SPIDER4
71
72   Naux= N−[Sof∗cos(phi)∗sin(psi);Sof∗sin(phi)∗sin(psi);Sof∗
         cos(phi)+Aof];
73
74
```

```
75  % Position of the spindle
76  S=[Naux;phi;psi];
```

3.5.3 Velocity relationship and Jacobian analysis

The Jacobian matrices for SPIDER4 are generated considering only the parallel structure mechanism, since the wrist mechanism's velocities are the same for the Cartesian and joint spaces. The Jacobian matrices allow us to map the actuated joint velocities to the velocities of the traveling plate. Having in mind that $||^oC_i - {}^o B_i||^2 = BC_{ix}^2 + BC_{iy}^2 + BC_{iz}^2$ Eq. (3.65) can be rewritten as:

$$
\begin{aligned}
x_n^2 &- 2Lx \cos\alpha_i \cos q_i + L^2 \cos^2\alpha_i \cos^2 q_i - 2aL \cos\alpha_i \cos q_i + a^2 \\
&+ y_n^2 - 2Ly \sin\alpha_i \cos q_i + L^2 \sin^2\alpha_i \cos^2 q_i - 2bL \sin\alpha_i \cos q_i + b^2 \quad (3.75) \\
&+ z_n^2 + 2Lz \sin q_i + L^2 \sin^2 q_i = l
\end{aligned}
$$

where $a = R_p \cos\alpha_i - R_b \cos\alpha_i$ and $b = R_p \sin\alpha_i - R_b \sin\alpha_i$. The time derivative of (3.76) yields the general equation of velocities' relationship expressed in this form:

$$
\begin{aligned}
&((R_p - R_b) \cos\alpha_i + x - L \cos\alpha_i \cos q_i)\dot{x} \\
&+ ((R_p - R_b) \sin\alpha_i + y - L \sin\alpha_i \cos q_i)\dot{y} + (z + L \sin q_i)\dot{z} \quad (3.76) \\
&= -L((x \cos\alpha_i + y \sin\alpha_i + R_p - R_b) \sin q_i + z \cos q_i)\dot{q}_i \quad \forall i = 1, ..., 4
\end{aligned}
$$

The terms of the previous expression are grouped with respect to the velocities in Cartesian space and in joint space in order to obtain the Jacobian matrices $\mathbf{J}_x \in \mathbb{R}^{4\times 3}$, and $\mathbf{J}_q \in \mathbb{R}^{4\times 4}$. These matrices can be expressed by:

$$
\mathbf{J}_x = \begin{bmatrix}
R\cos\alpha_1 + x_n - L\cos\alpha_1 \cos q_1 & R\sin\alpha_1 + y_n - L\sin\alpha_1 \cos q_1 & z_n + L\sin q_1 \\
R\cos\alpha_2 + x_n - L\cos\alpha_2 \cos q_2 & R\sin\alpha_2 + y_n - L\sin\alpha_2 \cos q_2 & z_n + L\sin q_2 \\
R\cos\alpha_3 + x_n - L\cos\alpha_3 \cos q_3 & R\sin\alpha_3 + y_n - L\sin\alpha_3 \cos q_3 & z_n + L\sin q_3 \\
R\cos\alpha_4 + x_n - L\cos\alpha_4 \cos q_3 & R\sin\alpha_4 + y_n - L\sin\alpha_4 \cos q_4 & z_n + L\sin q_4
\end{bmatrix}
$$
$$(3.77)$$

$$
\mathbf{J}_q = \begin{bmatrix}
Jq_{11} & 0 & 0 & 0 \\
0 & Jq_{22} & 0 & 0 \\
0 & 0 & Jq_{33} & 0 \\
0 & 0 & 0 & Jq_{44}
\end{bmatrix}
$$
$$(3.78)$$

Where

$$Jq_{11} = -L((x_n \cos \alpha_1 + y_n \sin \alpha_1 + R) \sin q_1 + z_n \cos q_1)$$
$$Jq_{22} = -L((x_n \cos \alpha_2 + y_n \sin \alpha_2 + R) \sin q_2 + z_n \cos q_2)$$
$$Jq_{33} = -L((x_n \cos \alpha_3 + y_n \sin \alpha_3 + R) \sin q_3 + z_n \cos q_3)$$
$$Jq_{44} = -L((x_n \cos \alpha_4 + y_n \sin \alpha_4 + R) \sin q_4 + z_n \cos q_4)$$

Therefore, the general Jacobian matrix for SPIDER4 RA-PKM may be written as follows:

$$\mathbf{J} = \mathbf{J}_q^{-1}\mathbf{J}_x \in \mathbb{R}^{4\times 3} \qquad (3.79)$$

The joint velocities of SPIDER4 RA-PKM, namely $\dot{\mathbf{q}} \in \mathbb{R}^{4\times 1}$ can be computed from the Cartesian velocities of the traveling plate $\dot{\mathbf{N}} \in \mathbb{R}^{3\times 1}$ through the general Jacobian matrix $\mathbf{J} \in \mathbb{R}^{4\times 3}$ in this way:

$$\dot{\mathbf{q}} = \mathbf{J}^o\dot{\mathbf{N}}_N \qquad (3.80)$$

However, to find the Cartesian velocities of the traveling plate in function of the joint velocities it is necessary to compute the pseudoinverse Jacobian matrix $\mathbf{H}$ since $\mathbf{J}$ is a non-square matrix. The pseudoinverse Jacobian matrix for SPIDER4 is then defined as:

$$\mathbf{H} = (\mathbf{J}^T\mathbf{J})^{-1}\mathbf{J}^T \in \mathbb{R}^{3\times 4} \qquad (3.81)$$

Consequently, the Cartesian velocities of the parallel structure of SPIDER4 can be computed through the following expression:

$$^o\dot{\mathbf{N}}_N = \mathbf{H}\dot{\mathbf{q}} \qquad (3.82)$$

ALGORITHM 3.6 (Programming code in MATLAB of the Jacobian matrix of SPIDER4 RA-PKM).
For a better understanding of the equations above, we provide the programming code in MATLAB.

```
1  function J = jac_Spider4(q,N)
2
3  q1=q(1);
4  q2=q(2);
5  q3=q(3);
6  q4=q(3);
7
8  xn=N(1);
9  yn=N(2);
```

```
10   zn=N(3);
11
12   %%% Kinematic parameters of SPIDER4 RA-PKM
13   L=0.535;
14   l=1.10;
15   Rb=0.4;
16   Rp=0.26;
17   Aof=0.198375;
18   Sof=0.135;
19   R=Rp-Rb;
20
21   %%% Angles between each kinematic chain
22   alpha1=7*pi/4;
23   alpha2=5*pi/4;
24   alpha3=3*pi/4;
25   alpha4=pi/4;
26
27   jx11=R*cos(alpha1)+xn-L*cos(alpha1)*cos(q1);
28   jx12=R*sin(alpha1)+yn-L*sin(alpha1)*cos(q1);
29   jx13=zn+L*sin(q1);
30
31   jx21=R*cos(alpha2)+xn-L*cos(alpha2)*cos(q2);
32   jx22=R*sin(alpha2)+yn-L*sin(alpha2)*cos(q2);
33   jx23=zn+L*sin(q2);
34
35   jx31=R*cos(alpha3)+xn-L*cos(alpha3)*cos(q3);
36   jx32=R*sin(alpha3)+yn-L*sin(alpha3)*cos(q3);
37   jx33=zn+L*sin(q3);
38
39   jx41=R*cos(alpha4)+xn-L*cos(alpha4)*cos(q4);
40   jx42=R*sin(alpha4)+yn-L*sin(alpha4)*cos(q4);
41   jx43=zn+L*sin(q4);
42
43   Jx=[jx11 jx12 jx13;jx21 jx22 jx23;jx31 jx32 jx33;jx41
        jx42 jx43];
44
45   jq11=-L*(sin(q1)*(xn*cos(alpha1)+yn*sin(alpha1)+R)+zn*cos
        (q1));
46   jq22=-L*(sin(q2)*(xn*cos(alpha2)+yn*sin(alpha2)+R)+zn*cos
        (q2));
47   jq33=-L*(sin(q3)*(xn*cos(alpha3)+yn*sin(alpha3)+R)+zn*cos
        (q3));
```

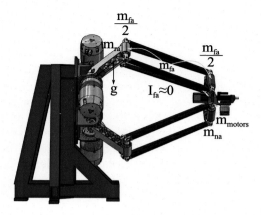

FIGURE 3.15 Illustration of dynamic model simplifications considered for SPIDER4 RA-PKM.

```
48  jq44=-L*(sin(q4)*(xn*cos(alpha4)+yn*sin(alpha4)+R)+zn*cos
       (q4));

49
50  Jq=[jq11 0 0 0;0 jq22 0 0;0 0 jq33 0;0 0 0 jq44];

51
52  J = inv(Jq)*Jx;
```

3.5.4 Inverse dynamic model of the delta-like positioning mechanism

The IDM of delta-like positioning device of SPIDER4 is obtained using the same modeling simplifications described in Chapter 2 and used for the Delta robot [84]. Reminding the equation that describes the dynamic model of a delta-like PKM (3.12)

$$\mathbf{\Gamma}(t) = \mathbf{\Gamma}_{sys} + \mathbf{\Gamma}_{tp}$$

Where $\mathbf{\Gamma}_{sys} \in \mathbb{R}^{4 \times 1}$ is the vector of the forces/torques contributions of the actuators and the rear-arms plus a half mass of the forearms, whereas $\mathbf{\Gamma}_{tp} \in \mathbb{R}^{4 \times 1}$ represents the forces/torques contribution of the mass of the traveling plate plus the other half mass of the forearms. The IDM for the parallel structure of SPIDER4 has notable differences compared to the Delta robot regarding the gravity effect on the kinematic chains since they are oriented horizontally whereas those of Delta are oriented vertically. Fig. 3.15 illustrates the modeling simplifications considered for SPIDER4. Let us first start with the computation of $\mathbf{\Gamma}_{tp}$. Considering that the traveling plate is performing only translation movements, it would be possible to avoid the calculation of moments over it. Therefore, we can describe the traveling plate dynamics by NE formulation as follows:

$$\mathbf{M}_t({}^o\ddot{\mathbf{N}}_N - \mathbf{g}) = 0 \tag{3.83}$$

Where, ${}^o\ddot{\mathbf{N}}_N \in \mathbb{R}^{3\times 1}$ denotes the Cartesian acceleration of the traveling plate, and $\mathbf{g} \in \mathbb{R}^{3\times 1}$ is the gravity acceleration vector denoted by $\mathbf{g} = \begin{bmatrix} 0 & g & 0 \end{bmatrix}^T$ with $g = 9.81 \ m/s^2$. $\mathbf{M}_t = \text{diag}([m_t]) \in \mathbb{R}^{3\times 3}$ is a diagonal matrix that contains the mass of the traveling plate, the mass of the actuators located on the traveling plate, and the contributions of the halves of the mass of the forearms to the traveling plate. The elements of this matrix are:

$$m_t = m_{tp} + m_{namotors} + 4\frac{m_{fa}}{2} \tag{3.84}$$

Where m_{tp} is the mass of the traveling plate, and $m_{namotors}$ is the mass of the three motors located at the traveling plate and their coupling parts. The motors located at the traveling plate include: the motor generating the movement around ϕ, the motor producing the movement over ψ, and the spindle motor. It is worth noting that the gravity acceleration is along the y-axis instead of the z-axis due to the orientation of the parallel structure of the robot. Therefore, the gravitational force acting on the traveling plate can be expressed as follows:

$$\mathbf{G}_t = -\mathbf{M}_t\mathbf{g} \tag{3.85}$$

The inertial forces acting on the traveling plate owing to the Cartesian acceleration ${}^o\ddot{\mathbf{N}}_N \in \mathbb{R}^{3\times 1}$ are expressed by:

$$\mathbf{F}_t = \mathbf{M}_t{}^o\ddot{\mathbf{N}}_N \tag{3.86}$$

The contributions of $\mathbf{G}_t$ and $\mathbf{F}_t$ in joint space can be calculated using the pseudoinverse Jacobian matrix $\mathbf{H}(\mathbf{q}, {}^o\mathbf{N}_N) \in \mathbb{R}^{3\times 4}$. These contributions are expressed by the following relationship:

$$\mathbf{\Gamma}_{G_t} = -\mathbf{H}^T\mathbf{M}_t\mathbf{g} \tag{3.87}$$

$$\mathbf{\Gamma}_{F_t} = \mathbf{H}^T\mathbf{M}_t{}^o\ddot{\mathbf{N}}_N \tag{3.88}$$

Therefore, $\mathbf{\Gamma}_{tp}$ is given as follows:

$$\mathbf{\Gamma}_{tp} = \mathbf{\Gamma}_{G_{tp}} + \mathbf{\Gamma}_{F_{tp}} = \mathbf{H}^T\mathbf{M}_t({}^o\ddot{\mathbf{N}}_N + \mathbf{g}) \tag{3.89}$$

Having the dynamic equations of the traveling plate, the next step consists in identifying the dynamics of the acting mechanism which is formed by the set of the actuators, the rear-arms, and the forearms. Similarly, as the traveling plate, we should compute the torques produced by the joint acceleration and gravity acceleration. The torques produced by joint acceleration on the kinematic chains $\boldsymbol{\Gamma}_{F_{rf}} \in \mathbb{R}^{4\times1}$ are expressed as follows:

$$\boldsymbol{\Gamma}_{F_{rf}} = \mathbf{I}_t \ddot{\mathbf{q}} \tag{3.90}$$

Where $\mathbf{I}_t \in \mathbb{R}^{4\times4}$ is a diagonal matrix whose elements are expressed by:

$$I_t = I_{act} + I_{ra} + \frac{L_i^2 m_{fa}}{2} \tag{3.91}$$

where I_{act}, I_{ra} represent inertias of the actuators and the rear-arms, respectively. The term $\frac{L_i^2 m_{fa}}{2}$ corresponds to the inertial contribution of the half mass of the forearms, where L is the rear-arms' length. In order to obtain the torques produced by the gravity acceleration acting on the kinematic chains, we need to take into account the orientation of the manipulator since the gravity force affects each one of the kinematic chains differently. The gravity effect over the limbs of SPIDER4 is directly related to its inclination w.r.t. the z-axis with a fixed angular orientation α_i as illustrated in Fig. 3.11. We may define the vector of torques produced by the gravitational forces acting on the kinematic chains set in following form:

$$\boldsymbol{\Gamma}_{G_{rf}} = -\mathbf{M}_{rf} g \cos(\mathbf{q}) \tag{3.92}$$

Where $\cos(\mathbf{q}) = \begin{bmatrix} \cos(q_1) & \cos(q_2) & \cos(q_3) & \cos(q_4) \end{bmatrix}^T$ the matrix $\mathbf{M}_{fra} \in \mathbb{R}^{4\times4}$ is composed as follows:

$$\mathbf{M}_{rf} = \begin{bmatrix} m_{fra}\sin(\alpha_1) & 0 & 0 & 0 \\ 0 & m_{fra}\sin(\alpha_2) & 0 & 0 \\ 0 & 0 & m_{fra}\sin(\alpha_3) & 0 \\ 0 & 0 & 0 & m_{fra}\sin(\alpha_4) \end{bmatrix} \tag{3.93}$$

In which, $m_{rf} = m_{ra}L_c + \frac{m_{fa}L}{2}$, L_c is the distance from the revolute joint of each rear-arm to its center of mass. By substituting (3.87), (3.88), (3.90), and (3.92) into (3.12), the following expression is obtained:

$$\boldsymbol{\Gamma} = \mathbf{I}_t \ddot{\mathbf{q}} + \mathbf{M}_{rf} g \cos(\mathbf{q}) + \mathbf{H}^T \mathbf{M}_t (^o\ddot{\mathbf{N}}_N + \mathbf{g}) \tag{3.94}$$

Eq. (3.94) is a representation of the IDM combining both Cartesian and joint space variables. However, in order to express the dynamic model of the delta-like positioning device of SPIDER4 only in the joint space, we have to use the following relationship, based on the pseudoinverse Jacobian matrix and its derivative:

$$^o\ddot{\mathbf{N}}_N = \mathbf{H}\ddot{\mathbf{q}} + \dot{\mathbf{H}}\dot{\mathbf{q}} \tag{3.95}$$

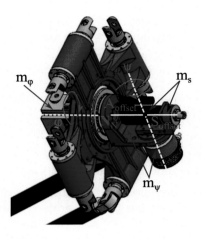

FIGURE 3.16   Close view of the SPIDER4 wrist.

By rearranging the terms, the simplified IDM of the delta-like positioning mechanism of SPIDER4 can be expressed in terms of the redundant joint coordinates $\mathbf{q}$ as follows:

$$\mathbf{M}(\mathbf{q})\ddot{\mathbf{q}} + \mathbf{C}(\mathbf{q}, \dot{\mathbf{q}})\dot{\mathbf{q}} + \mathbf{G}(\mathbf{q}) = \mathbf{\Gamma}(t) \qquad (3.96)$$

Where:

- $\mathbf{M}(\mathbf{q}) = \mathbf{I}_t + \mathbf{H}^T \mathbf{M}_t \mathbf{H}$
- $\mathbf{C}(\mathbf{q}, \dot{\mathbf{q}}) = \mathbf{H}^T \mathbf{M}_t \dot{\mathbf{H}}$
- $\mathbf{G}(\mathbf{q}) = \mathbf{H}^T \mathbf{M}_t \mathbf{g} + \mathbf{M}_{rf}\, g \cos(\mathbf{q})$

Where $\mathbf{M}(\mathbf{q}) \in \mathbb{R}^{4\times4}$ denotes the inertia matrix of the delta-like positioning device of SPIDER4, $\mathbf{C}(\mathbf{q}, \dot{\mathbf{q}}) \in \mathbb{R}^{4\times4}$ is the Coriolis/Centripetal forces matrix satisfying the skew-symmetry property; $\mathbf{G}(\mathbf{q}) \in \mathbb{R}^{4\times1}$ is the vector of gravitational forces and torques being continuous and bounded.

3.5.5 Inverse dynamic model of the wrist

In the previous section, a simplified IDM for the delta-like positioning device of SPIDER4 was derived. This section aims to compute the IDM of the wrist in order to establish the IDM of the whole SPIDER4 RA-PKM. The wrist is a serial mechanism formed by two links actuated by two rotational motors. The IDM of this part of the robot is computed using the Euler-Lagrange formulation. Thus, we need first to obtain the forward kinematic model of the wrist, which is obtained considering Eq. (3.3). Fig. 3.16 illustrates the dynamic and kinematic parameters involved in this analysis. The forward kinematic model of the wrist is expressed in the traveling plate reference frame $N - x_n, y_n, z_n$ as follows:

$$^N\mathbf{W}_W = \begin{bmatrix} x_w \\ y_w \\ z_w \end{bmatrix} = \begin{bmatrix} -S_{offset}\cos(\phi)\sin(\psi) \\ -S_{offset}\sin(\phi)\sin(\psi) \\ -S_{offset}\cos(\psi) + A_{offset} \end{bmatrix} \quad (3.97)$$

The Euler-Lagrange formulation needs the equation of the kinetic and potential energy of the wrist. Developing the derivative of (3.97), the kinetic energy of the wrist can be expressed as:

$$\mathbf{K}_w = \frac{1}{2}m_s^N\dot{\mathbf{W}}_W^{TN}\dot{\mathbf{W}}_W + \frac{1}{2}I_{m\phi}\dot{\phi} \quad (3.98)$$

Where m_s is the mass of the spindle motor and its coupling parts and $I_{m\phi}$ is the inertia produced by the mass of the second motor mounted on the traveling plate and its coupling parts. The Potential energy of the wrist can be expressed in the following form, considering that the gravity acceleration is over the y-axis.

$$\mathbf{U}_w = -m_s g S_{offset}\sin(\phi)\sin(\psi) \quad (3.99)$$

Hence, the Lagrangian expression is established in function of $\boldsymbol{\xi} = \begin{bmatrix} \phi & \psi \end{bmatrix}^T$, as follows [116]:

$$\mathcal{L}(\boldsymbol{\xi}, \dot{\boldsymbol{\xi}}) = \mathbf{K}_w(\boldsymbol{\xi}, \dot{\boldsymbol{\xi}}) - \mathbf{U}_w(\boldsymbol{\xi}) \quad (3.100)$$

The Lagrangian function (3.100) is evaluated in the Euler-Lagrange dynamic equation, to compute the IDM of the wrist, leading to:

$$\frac{d}{dt}\left[\frac{\partial \mathcal{L}(\boldsymbol{\xi}, \dot{\boldsymbol{\xi}})}{\partial \dot{\boldsymbol{\xi}}}\right] - \left[\frac{\partial \mathcal{L}(\boldsymbol{\xi}, \dot{\boldsymbol{\xi}})}{\partial \boldsymbol{\xi}}\right] = \boldsymbol{\Gamma}_w \quad (3.101)$$

Being $\boldsymbol{\Gamma}_w \in \mathbb{R}^{2\times 1}$ the torque vector generated by the wrist actuators. Developing Eq. (3.101); the IDM of the wrist is formulated in function of its joint coordinates as follows:

$$\mathbf{M}_w(\boldsymbol{\xi})\ddot{\boldsymbol{\xi}} + \mathbf{C}_w(\boldsymbol{\xi}, \dot{\boldsymbol{\xi}})\dot{\boldsymbol{\xi}} + \mathbf{G}_w(\boldsymbol{\xi}) = \boldsymbol{\Gamma}_w \quad (3.102)$$

being

$$\mathbf{M}_w(\boldsymbol{\xi}) = \begin{bmatrix} m_s S_{offset}^2 \sin\psi + I_{m\phi} & 0 \\ 0 & m_s S_{offset}^2 \end{bmatrix}$$

$$\mathbf{C}_w(\boldsymbol{\xi}, \dot{\boldsymbol{\xi}}) = \begin{bmatrix} 0.5 m_s S_{offset} \cos\psi\dot{\phi} & 0 \\ 0 & 0 \end{bmatrix}$$

$$\mathbf{G}_w(\boldsymbol{\xi}) = \begin{bmatrix} -m_s\, g\, S_{offset} \sin \psi \cos \phi \\ -m_s\, g\, S_{offset} \sin \phi \cos \psi \end{bmatrix}$$

where $\mathbf{M}_w(\boldsymbol{\xi}) \in \mathbb{R}^{2\times2}$ denotes the inertia matrix of the wrist, $\mathbf{C}_w(\boldsymbol{\xi}, \dot{\boldsymbol{\xi}}) \in \mathbb{R}^{2\times2}$ is the matrix of Coriolis and centrifugal effects, and $\mathbf{G}_w(\boldsymbol{\xi}) \in \mathbb{R}^{2\times1}$ is the gravity vector of the wrist.

3.5.6 Inverse dynamic model of SPIDER4 RA-PKM

The whole inverse dynamic model of SPIDER4 can be expressed in terms of the generalized coordinates, as a result of the combination of the IDM of the positioning delta-like device (3.94), and the inverse dynamic model of the wrist (3.102), leading to:

$$\mathbf{M}(\mathbf{Q})\ddot{\mathbf{Q}} + \mathbf{C}(\mathbf{Q}, \dot{\mathbf{Q}})\dot{\mathbf{Q}} + \mathbf{G}(\mathbf{Q}) = \boldsymbol{\Gamma}_T(t) \tag{3.103}$$

being:

$$\mathbf{M}(\mathbf{Q}) = \begin{bmatrix} \mathbf{M}(\mathbf{q})_{[4\times4]} & \mathbf{0}_{[2\times4]} \\ \mathbf{0}_{[2\times4]} & \mathbf{M}_w(\boldsymbol{\xi})_{[2\times2]} \end{bmatrix}$$

$$\mathbf{C}(\mathbf{Q}, \dot{\mathbf{Q}}) = \begin{bmatrix} \mathbf{C}(\mathbf{q}, \dot{\mathbf{q}})_{[4\times4]} & \mathbf{0}_{[2\times4]} \\ \mathbf{0}_{[2\times4]} & \mathbf{C}_w(\boldsymbol{\xi}, \dot{\boldsymbol{\xi}})_{[2\times2]} \end{bmatrix}$$

$$\mathbf{G}(\mathbf{Q}) = \begin{bmatrix} \mathbf{G}(\mathbf{q})_{[4\times1]} \\ \mathbf{G}_w(\boldsymbol{\xi})_{[2\times1]} \end{bmatrix}; \quad \boldsymbol{\Gamma}_T = \begin{bmatrix} \boldsymbol{\Gamma}_{[4\times1]} \\ \boldsymbol{\Gamma}_{w[2\times1]} \end{bmatrix}; \quad \mathbf{Q} = \begin{bmatrix} \mathbf{q}_{[4\times1]} \\ \boldsymbol{\xi}_{[2\times1]} \end{bmatrix}$$

where $\mathbf{M}(\mathbf{Q}) \in \mathbb{R}^{6\times6}$ denotes the inertia matrix of SPIDER4, $\mathbf{C}(\mathbf{Q}, \dot{\mathbf{Q}}) \in \mathbb{R}^{6\times6}$ is the matrix of Coriolis and centrifugal effects of SPIDER4, $\mathbf{G}(\mathbf{Q}) \in \mathbb{R}^{6\times1}$ is the gravity vector and, $\boldsymbol{\Gamma}_T \in \mathbb{R}^{6\times1}$ is the torque vector of all the actuators of SPIDER4.

The dynamic parameters of SPIDER4 as the masses of the rear-arms, forearms, the traveling plate, coupling parts, and rear-arms inertia were calculated through the material assignation functionality in SolidWorks CAD software. The other dynamic parameters as the inertias and masses of the motors were obtained from the datasheets of the manufacturers of the actuators. The values of these dynamic parameters are summarized in Table 3.4.

NOTE: The IDM of (3.96) is considered for the proposed control solutions instead of (3.103), since the delta-like positioning device and the wrist mechanisms are controlled independently. The robot's low-level software allows us to modify only the control scheme for the delta-like mechanism. In next section, the modeling of the 3-DOF Delta PKM will be detailed.

TABLE 3.4 Summary of the dynamic parameters of SPIDER4.

Parameter	Description	Value
m_{tp}	Traveling plate mass	22.76 kg
$m_{namotors}$	Mass of the three motors mounted on the traveling plate	19.5 kg
m_{ra}	Rear-arm mass	17.6 kg
m_{fa}	Forearm mass	4.64 kg
I_{arm}	Rear-arm inertia	1.69 kg m^2
I_{act}	Inertia of one of the principal actuators	0.00223 kg m^2
m_s	Mass of the spindle motor	3.2 kg
m_ψ	Motor mass that regulates movement in ψ	5.1 kg
m_ϕ	Motor mass that regulates movement in ϕ	11.2 kg

3.6 The actuation redundancy issue on SPIDER4 RA-PKM

In Chapter 1, the different challenges involved in the control of PKMs were explained. One of them is the actuation redundancy which is present on Parallel robots having more actuators than DOF [81]. Such is the case of the SPIDER4, which has 5-DOF (3T-2T), where the positioning device develops the 3T-DOF using four actuators located on the fixed base. We mentioned that robotics systems with this type of configuration have advantages compared with their non-redundant counterparts, such as higher accuracy and improved stiffness. Furthermore, actuation redundancy can also lead to singularity-free large workspaces. However, this configuration leads to the generation of internal forces that may create pre-stress in the mechanism without operational motions, and it can damage the mechanical structure of the robot. According to [67], internal forces may be caused by geometric uncertainties and may be amplified by the use of decentralized control techniques. This may lead to uncoordinated control of the individual actuators since such a control law does not account for kinematic constraints. In order to avoid such issue, the use of a projection matrix can be used, which is based on the pseudo inverse Jacobian matrix evaluated with the desired variables $\mathbf{H}(\mathbf{q_d},{}^o\mathbf{N}_{Nd}) \in \mathbb{R}^{m \times n}$. The projection operator is defined by:

$$\mathbf{R}_H = (\mathbf{H}^+)^T \mathbf{H}^T \qquad (3.104)$$

The projection matrix $\mathbf{R}_H$ eliminates the control inputs in the null-space of $\mathbf{H}^T$. Hence, all control inputs applied to the manipulator have to be 'regularized' using this projection matrix as follows:

$$\mathbf{\Gamma}^* = \mathbf{R}_H \mathbf{\Gamma} \qquad (3.105)$$

Where $\mathbf{\Gamma}$ denotes the torques' vector generated by the control scheme. Therefore, all torques generated by the presented control solutions for SPIDER4, must be regularized before sending them to the manipulator.

3.7 Conclusion

In this chapter, two PKMs were introduced, the new machining PKM called SPIDER4, which was built between LIRMM and Tecnalia, and a Delta PKM designed and built at the Polytechnic University of Tulancingo intended to be used in P&P tasks. The kinematic and dynamic models of both prototypes were also developed within this book. For SPIDER4 positioning device and the Delta PKM, the closed-loop method was used to obtain the IKM. Besides, transformations matrices have been employed to compute the IKM of the wrist of SPIDER4. The IKM of the delta-like positioning device and the wrist were coupled to get the whole IKM of SPIDER4. The FKM for both PKMs has been calculated through a virtual spheres' intersection algorithm, which provides the traveling-plate position when the robot motors' joint position is known. The expression of velocities was achieved thanks to the Jacobian matrices. These matrices were derived from the closed-loop equation established in the kinematic analysis to obtain the IKMs. Two significant simplifications were considered in the IDM development for the delta-like positioning device of SPIDER4 and the Delta PKM. The first simplification consists of neglecting the inertia of the forearms and dividing their mass into two parts. One part is added to the end of the rear-arm, and the other one is added to the traveling plate. The second simplification neglects the frictional forces in the joints of the PKMs since it is complicated to obtain an accurate friction model in highly nonlinear systems as these kinds of PKMs. Moreover, in the case of SPIDER4, the IDM of the wrist was calculated using the Euler-Lagrange formulation. Both IDMs of the wrist and the positioning part were joined together to get the whole IDM of SPIDER4. Nevertheless, we will consider only the IDM of the delta-like device for control purposes since the software of SPIDER4 do not allow low-level control of the wrist mechanism.

CHAPTER

4

Proposed robust control solutions

4.1 Introduction

Control design of PKMs is an important research area of robotics because it determines the robotic system's motion. As mentioned in Chapter 1, parallel robots, due to their kinematic construction, have various
desirable features such as: improved stiffness, equitable distribution of the
payload in their set of linkages, high repeatability, potential accuracy, and,
in the case of RA-PKMs, singularity avoidance, and improved dynamics.
However, the control of parallel robots represents a significant challenge
due to the presence of different types of uncertainties, the highly nonlinear dynamics, and, in the case of RA-PKMs, the problem of internal forces
generation. To overcome these issues, advanced nonlinear control techniques may be a good solution to obtain the highest potential that PKMs
can offer. Aware of the importance of designing advanced control schemes
for Parallel robots, in this chapter, we will describe in detail two proposed
control solutions for PKMs. The proposed controllers aim to keep the joint
tracking errors as small as possible under the environment in which the
robot is intended to operate. The first control solution aims to validate the
proposed IDM of SPIDER4 experimentally, the second one is designed to
deal with the problem of machining. Although these control schemes were
designed for machining and P&P tasks with PKMs, it is also possible to implement them in any other types of robot manipulators. Before describing
these proposed control schemes in detail, let us present a background of
the (Robust Integral of the Sign of the Error) RISE control scheme, which
is the core of the proposed control solutions.

Modeling and Nonlinear Robust Control of Delta-Like Parallel Kinematic Manipulators
https://doi.org/10.1016/B978-0-32-396101-1.00011-X

4.2 Background on RISE feedback control

RISE is a kind of continuous feedback controller proved to ensure semi-global asymptotic stability even in the presence of a large class of uncertainties and nonlinearities [133]. Must nonlinear systems are well known by the presence of non-modeled phenomena such as friction dynamics, or the inaccurate knowledge of their dynamic parameters. Indeed, in robotic manipulators, some of such dynamic parameters are also time-varying, such as the payload. Therefore, based on basic assumptions on the system to be controlled as the equation system structure, the controller can partially compensate for such nonlinearities and uncertainties with an appropriate feedback gain selection. For the mentioned advantages, RISE control has already been tested on many robotic devices, such as underactuated robotic systems [62], [60], exoskeletons [114], Hard disks [123], [124], [122], and of course parallel robots [64]. We can detail some examples of application of the RISE control in PKMs. For instance, [16], reports the first implementation of the original RISE control on a 3-DOF Delta robot. In the mentioned paper, the authors also enhance the RISE feedback controller with and adaptive feedforward term reducing the tracking errors. A similar RISE control with an adaptive feedforward was implemented on a 3-DOF RA-PKM called Dual-V [9]. Another interesting extension of RISE control was introduced in [106] and [105], where it was proposed to replace the constant feedback gains with nonlinear time-varying ones endowing the controller with better robustness against external perturbations. Moreover, RISE control has been complemented with neural networks for the estimation of system dynamics. For instance, in [49] RISE control was complemented with a B-Spline neural network. The objective of this neural network was to estimate the dynamics of the robot and to integrate it into the control loop as a feedforward term.

Having detailed some application examples, let us describe the formulation of this control scheme.

Let us consider the following p^{th} order MIMO nonlinear system

$$\mathbf{M}(\mathbf{X})\mathbf{X}^{(p)} + \mathbf{F}(\mathbf{X}) = \mathbf{U}(t) \tag{4.1}$$

In which, $\mathbf{X}(t) = [\mathbf{x}(t) \quad \dot{\mathbf{x}}(t) \quad , ..., \quad \mathbf{x}^{p-1}(t)]^T \in \mathbb{R}^n$ represents the system states, $\mathbf{U}(t) \in \mathbb{R}^n$ denotes the control input, and $\mathbf{M}(\mathbf{X}) \in \mathbb{R}^{n \times n}$, $\mathbf{F}(\mathbf{X}) \in \mathbb{R}^n$ are uncertain nonlinear functions. Based on the system states and the desired trajectories one can define the output (position) tracking error as follows:

$$\mathbf{e}_1 = \mathbf{x}_d - \mathbf{x} \tag{4.2}$$

Where $\mathbf{x}_d \in \mathbb{R}^n$ represents the desired trajectory. The control objective is to guarantee that $\mathbf{e}_1(t) \to 0$ as $t \to \infty$.

Definition 1. The nonlinear functions $\mathbf{M}(.)$, $\mathbf{F}(.)$ are second-order differentiable and bounded.

Definition 2. The uncertain nonlinear function $\mathbf{M}(.) \in \mathbb{R}^{n \times n}$ is a symmetric positive-definite matrix satisfying the following inequality for all $\boldsymbol{\zeta} \in \mathbb{R}^n$:

$$\underline{m}||\boldsymbol{\zeta}||^2 \leq \boldsymbol{\zeta}^T \mathbf{M}(.) \leq \overline{m}(\mathbf{x})||\boldsymbol{\zeta}||^2 \tag{4.3}$$

Where $\underline{m}$ is a positive constant, $\overline{m}(\mathbf{x})$ is a positive non-decreasing function, and $||.||$ represents the standard Euclidean norm.

Definition 3. The desired trajectory $\mathbf{x}_d$ is continuously differentiable with respect to time until the $(n + 2)$ derivative.

4.2.1 Control law

Lets us now consider the following auxiliary error signals before introducing the control law:

$$\begin{aligned}
\mathbf{e}_2(t) &= \dot{\mathbf{e}}_1(t) + \mathbf{e}_1(t) \\
\mathbf{e}_3(t) &= \dot{\mathbf{e}}_2(t) + \mathbf{e}_2(t) + \mathbf{e}_1(t) \\
\mathbf{e}_4(t) &= \dot{\mathbf{e}}_3(t) + \mathbf{e}_3(t) + \mathbf{e}_2(t) \\
&\vdots \\
\mathbf{e}_n(t) &= \dot{\mathbf{e}}_{n-1}(t) + \mathbf{e}_{n-1}(t) + \mathbf{e}_{n-2}(t)
\end{aligned} \tag{4.4}$$

According to the stability analysis presented in [133], the following control law is established to fulfill the control objective:

$$\mathbf{U}(t) = (\mathbf{K}_s + \mathbf{I})\mathbf{e}_n(t) - (\mathbf{K}_s + \mathbf{I})\mathbf{e}_n(0) + \int_0^t [(\mathbf{K}_s + \mathbf{I})\mathbf{\Lambda}\mathbf{e}_n(\tau) + \boldsymbol{\beta}\text{sgn}(\mathbf{e}_n(\tau))]d\tau \tag{4.5}$$

Where $\mathbf{K}_s$, $\mathbf{\Lambda}$, $\boldsymbol{\beta} \in \mathbb{R}^{n \times n}$ are positive-definite control gain matrices, and $\mathbf{I} \in \mathbb{R}^{n \times n}$ represents the identity matrix.

4.2.2 Application of standard RISE feedback control to PKMs

For a PKM with m-DOF and n actuators, Eq. (4.1) may be expressed as:

$$\mathbf{M}(\mathbf{q})\ddot{\mathbf{q}} + \mathbf{F}(\mathbf{q}, \dot{\mathbf{q}}) = \mathbf{\Gamma}(t) \tag{4.6}$$

Where $\mathbf{F}(\mathbf{q}, \dot{\mathbf{q}}) = \mathbf{C}(\mathbf{q}, \dot{\mathbf{q}})\dot{\mathbf{q}} + \mathbf{G}(\mathbf{q}) + \mathbf{f}(\mathbf{q}, \dot{\mathbf{q}}) + \mathbf{\Gamma}_d$, being $\mathbf{f}(\mathbf{q}, \dot{\mathbf{q}}) \in \mathbb{R}^n$ the vector containing the friction effects, and $\mathbf{\Gamma}_d \in \mathbb{R}^n$ a bounded vector of disturbances. The joint tracking error $\mathbf{e}_1(t) \in \mathbb{R}^n$ is then defined as:

$$\mathbf{e}_1(t) = \mathbf{q}_d(t) - \mathbf{q}(t) \tag{4.7}$$

The following filtered tracking errors are defined to further assist in the stability analysis of the resulting closed-loop system [93]:

$$\mathbf{e}_2(t) = \dot{\mathbf{e}}_1(t) + \mathbf{\Lambda}_1 \mathbf{e}_1(t) \tag{4.8}$$

$$\mathbf{r}(t) = \dot{\mathbf{e}}_2(t) + \mathbf{\Lambda}_2 \mathbf{e}_2(t) \tag{4.9}$$

Where $\mathbf{\Lambda}_1$ and $\mathbf{\Lambda}_2 \in \mathbb{R}^{n \times n}$, are two positive-definite diagonal gain matrices. The open-loop tracking error system can be obtained by multiplying (4.9) by $\mathbf{M}(\mathbf{q})$ yielding the following expression:

$$\mathbf{M}(\mathbf{q})\mathbf{r} = \mathbf{M}(\mathbf{q})(\dot{\mathbf{e}}_2 + \mathbf{\Lambda}_2 \mathbf{e}_2) \tag{4.10}$$

Making use of (4.9) and (4.8), the closed-loop tracking error system can be expressed as:

$$\mathbf{M}(\mathbf{q})\mathbf{r} = \mathbf{F}(\mathbf{q}, \dot{\mathbf{q}}) + \mathbf{M}(\mathbf{q})(\ddot{\mathbf{q}}_d + \mathbf{\Lambda}_1 \dot{\mathbf{e}}_1 + \mathbf{\Lambda}_2 \mathbf{e}_2) - \mathbf{\Gamma}_{RISE}(t) \tag{4.11}$$

In which, $\mathbf{\Gamma}_{RISE}(t) = \mathbf{\Gamma}(t)$. The control law (4.5) can be reformulated to be applied to robotic systems as follows:

$$\mathbf{\Gamma}_{RISE}(t) = (\mathbf{K}_s + \mathbf{I})\mathbf{e}_2(t) - (\mathbf{K}_s + \mathbf{I})\mathbf{e}_2(0)$$
$$+ \int_0^t [(\mathbf{K}_s + \mathbf{I})\mathbf{\Lambda}_2 \mathbf{e}_2(\tau) + \boldsymbol{\beta}\mathrm{sgn}(\mathbf{e}_2(\tau))]d\tau \tag{4.12}$$

To simplify the stability analysis of the resulting closed-loop system, one can compute the time derivative of (4.11), leading to the following equation:

$$\mathbf{M}(\mathbf{q})\dot{\mathbf{r}} = -\dot{\mathbf{M}}(\mathbf{q})(\mathbf{r} - \ddot{\mathbf{q}}_d) + \dot{\mathbf{F}}(\mathbf{q}, \dot{\mathbf{q}}) + \mathbf{M}(\mathbf{q})\dddot{\mathbf{q}}_d + \mathbf{\Lambda}_1(\mathbf{M}(\mathbf{q})\ddot{\mathbf{e}}_1 + \dot{\mathbf{M}}(\mathbf{q})\dot{\mathbf{e}}_1)$$
$$\mathbf{\Lambda}_2(\mathbf{M}(\mathbf{q})\dot{\mathbf{e}}_2 + \dot{\mathbf{M}}(\mathbf{q})\mathbf{e}_2) - \dot{\mathbf{\Gamma}}_{RISE} \tag{4.13}$$

This equation can be rewritten in the following form as stated in [134]

$$\mathbf{M}(\mathbf{q})\dot{\mathbf{r}} = -\frac{1}{2}\dot{\mathbf{M}}(\mathbf{q})\mathbf{r} + \mathbf{N}(\mathbf{e}_1, \mathbf{e}_2, \mathbf{r}, t) - \mathbf{e}_2(t) - \dot{\mathbf{\Gamma}}_{RISE}(t) \tag{4.14}$$

Where $\mathbf{N}(\mathbf{e}_1, \mathbf{e}_2, \mathbf{r}, t) \in \mathbb{R}^n$ is a nonlinear auxiliary function, containing the uncertain terms related to the robot's dynamic model, whose elements are defined as follows:

$$\mathbf{N}(\mathbf{e}_1, \mathbf{e}_2, \mathbf{r}, t) = \dot{\mathbf{M}}(\mathbf{q})\ddot{\mathbf{q}}_d + \mathbf{M}(\mathbf{q})\dddot{\mathbf{q}}_d + \dot{\mathbf{F}}(\mathbf{q}, \dot{\mathbf{q}}) + \mathbf{e}_2(t) - \frac{1}{2}\dot{\mathbf{M}}(\mathbf{q})\mathbf{r}$$
$$\mathbf{\Lambda}_1(\mathbf{M}(\mathbf{q})\ddot{\mathbf{e}}_1 + \dot{\mathbf{M}}(\mathbf{q})\dot{\mathbf{e}}_1) + \mathbf{\Lambda}_2(\mathbf{M}(\mathbf{q})\dot{\mathbf{e}}_2 + \dot{\mathbf{M}}(\mathbf{q})\mathbf{e}_2) \tag{4.15}$$

The time derivative of the standard RISE control is given as:

$$\dot{\mathbf{\Gamma}}_{RISE} = (\mathbf{K}_s + \mathbf{I})\mathbf{r} + \boldsymbol{\beta}\mathrm{sgn}(\mathbf{e}_2) \tag{4.16}$$

Let us now define the following auxiliary function $\mathbf{N}_d(\mathbf{q}_d, \dot{\mathbf{q}}_d, \ddot{\mathbf{q}}_d, t)\mathbb{R}^n$, to facilitate the subsequent stability analysis of the controller:

$$\mathbf{N}_d(\mathbf{q}_d, \dot{\mathbf{q}}_d, \ddot{\mathbf{q}}_d, t) = \mathbf{M}(\mathbf{q}_d)\dddot{\mathbf{q}}_d + \dot{\mathbf{M}}(\mathbf{q}_d)\ddot{\mathbf{q}}_d + \mathbf{C}(\mathbf{q}_d, \dot{\mathbf{q}}_d)\ddot{\mathbf{q}}_d \\ + \dot{\mathbf{C}}(\mathbf{q}_d, \dot{\mathbf{q}}_d)\dot{\mathbf{q}}_d + \dot{\mathbf{G}}(\mathbf{q}_d) + \dot{\mathbf{f}}(\mathbf{q}_d, \dot{\mathbf{q}}_d) \tag{4.17}$$

Adding and subtracting $\mathbf{N}_d(\mathbf{q}_d, \dot{\mathbf{q}}_d, \ddot{\mathbf{q}}_d, t)$ to the right-hand side of the closed-loop error system equation (4.14), the following mathematical expression is obtained:

$$\mathbf{M}(\mathbf{q})\dot{\mathbf{r}} = -\frac{1}{2}\dot{\mathbf{M}}(\mathbf{q})\mathbf{r} + \mathbf{N}_d + \tilde{\mathbf{N}} - \mathbf{e}_2(t) - \dot{\mathbf{\Gamma}}_{RISE}(t) \tag{4.18}$$

Where

$$\tilde{\mathbf{N}}(\mathbf{e}_1, \mathbf{e}_2, \mathbf{r}, t) = \mathbf{N}(\mathbf{e}_1, \mathbf{e}_2, \mathbf{r}, t) - \mathbf{N}_d(\mathbf{q}_d, \dot{\mathbf{q}}_d, \ddot{\mathbf{q}}_d, t) \tag{4.19}$$

The vector $\tilde{\mathbf{N}}$ can be upper bounded as follows using the Mean Value Theorem:

$$||\tilde{\mathbf{N}}|| \leq \rho(||\mathbf{z}||)||\mathbf{z}|| \tag{4.20}$$

In which, $|| \cdot ||$ represents the Euclidean norm, and $\mathbf{z}(t) \in \mathbb{R}^{3n}$ is an error vector defined as:

$$\mathbf{z}(t) = [\mathbf{e}_1 \quad \mathbf{e}_2 \quad \mathbf{r}]^T \tag{4.21}$$

The stability analysis for this control scheme was proved in [133], achieving the control objective introduced at the beginning of this section. The mentioned work will be considered for the analysis of the proposed control solutions detailed in further sections.

4.3 Control solution 1: A RISE controller with nominal feedforward

4.3.1 Motivation

Feedforward control is an open-loop scheme that compensates the system dynamics without needing information about the system states as the tracking error. For such reason, it requires precise knowledge about the dynamics of the system to be controlled. However, it is practically impossible to have accurate knowledge of the system parameters. Besides, the abundance of uncertainties makes the implementation of a feedforward control alone inappropriate because the tracking error will always be incremented as this control scheme does not offer any corrective action. Thus, it is necessary to integrate a feedback term guaranteeing that the tracking error goes to zero. In the literature, it has been demonstrated in several works that the addition of a feedforward term to a feedback controller can significantly enhance the overall performance over simple feedback control. We propose to combine the Standard RISE control as a robust feedback control to a nominal feedforward term. The main objective of this control solution is to demonstrate the effectiveness of the proposed IDM for Delta and SPIDER4 with their computed dynamic parameters. Considering the previous advantage mentioned regarding the addition of a feedforward term to a feedback controller, we can say that if the IDM of Delta and SPIDER4 and their parameters are sufficiently well approximated to the real dynamics, we will obtain better performance compared to the RISE control alone. However, in the opposite case, the resulting performance will be worse. Figs. 4.1 and 4.2 illustrate the block diagram of this control scheme applied to Delta PKM and SPIDER4 RA-PKM, respectively.

4.3.2 Proposed control law

Considering the advantages presented by RISE control and taking into account the benefits generated by the addition of model-based control terms to the control law, the following controller is proposed:

$$\mathbf{\Gamma}(t) = \mathbf{\Gamma}_{RISE} + \mathbf{\Gamma}_{FF} \qquad (4.22)$$

Where $\mathbf{\Gamma}_{RISE} \in \mathbb{R}^n$ is the same controller equation described in (4.12), and the nominal feedforward term $\mathbf{\Gamma}_{FF} \in \mathbb{R}^n$ is defined as follows

$$\mathbf{\Gamma}_{FF} = \mathbf{M}(\mathbf{q}_d)\ddot{\mathbf{q}}_d + \mathbf{C}(\mathbf{q}_d, \dot{\mathbf{q}}_d)\dot{\mathbf{q}}_d + \mathbf{G}(\mathbf{q}_d) \qquad (4.23)$$

4.3.2.1 Controller design

Substituting the proposed control law (4.22) into the dynamics of a PKM (4.6) the following closed-loop tracking error system is obtained

$$\mathbf{M}(\mathbf{q})\mathbf{r} = \mathbf{F}(\mathbf{q}, \dot{\mathbf{q}}) + \mathbf{M}(\mathbf{q})(\ddot{\mathbf{q}}_d + \mathbf{\Lambda}_1\dot{\mathbf{e}}_1 + \mathbf{\Lambda}_2\mathbf{e}_2) - \mathbf{\Gamma}_{RISE}(t) - \mathbf{\Gamma}_{FF}(t) \quad (4.24)$$

The first time-derivative of (4.24) results in the following expression:

$$\mathbf{M}(\mathbf{q})\dot{\mathbf{r}} = -\dot{\mathbf{M}}(\mathbf{q})(\mathbf{r} - \ddot{\mathbf{q}}_d) + \dot{\mathbf{F}}(\mathbf{q}, \dot{\mathbf{q}}) + \mathbf{M}(\mathbf{q})\dddot{\mathbf{q}}_d + \mathbf{\Lambda}_1(\mathbf{M}(\mathbf{q})\ddot{\mathbf{e}}_1 + \dot{\mathbf{M}}(\mathbf{q})\dot{\mathbf{e}}_1)$$
$$\mathbf{\Lambda}_2(\mathbf{M}(\mathbf{q})\dot{\mathbf{e}}_2 + \dot{\mathbf{M}}(\mathbf{q})\mathbf{e}_2) - \dot{\mathbf{\Gamma}}_{RISE} - \dot{\mathbf{\Gamma}}_{FF}$$
$$(4.25)$$

The previous equation is reformulated in the same way as was presented in subsection 4.2.1.

$$\mathbf{M}(\mathbf{q})\dot{\mathbf{r}} = -\frac{1}{2}\dot{\mathbf{M}}(\mathbf{q})\mathbf{r} + \mathbf{N}(\mathbf{e}_1, \mathbf{e}_2, \mathbf{r}, t) - \mathbf{e}_2(t) - \dot{\mathbf{\Gamma}}_{RISE}(t) - \dot{\mathbf{\Gamma}}_{FF}(t) \quad (4.26)$$

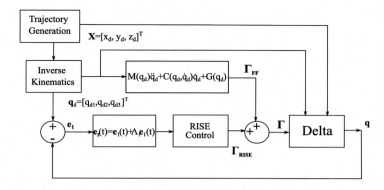

FIGURE 4.1   Block-diagram of proposed control solution 1 for Delta PKM.

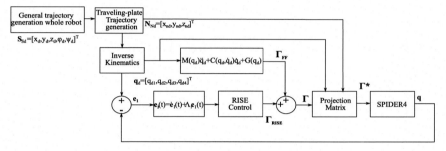

FIGURE 4.2   Block-diagram of proposed control solution 1 for the parallel device of SPIDER4 RA-PKM.

Where $\mathbf{N}(\mathbf{e}_1, \mathbf{e}_2, \mathbf{r}, t) \in \mathbb{R}^n$ was defined in (4.15). To facilitate the stability analysis of this control solution, we have to reformulate (4.26) as follows:

$$\mathbf{M}(\mathbf{q})\dot{\mathbf{r}} = -\frac{1}{2}\dot{\mathbf{M}}(\mathbf{q})\mathbf{r} + \mathbf{N}_d + \tilde{\mathbf{N}} - \mathbf{e}_2(t) - \dot{\mathbf{\Gamma}}_{RISE}(t) - \mathbf{N}_{FF} \qquad (4.27)$$

Where $\mathbf{N}_d \in \mathbb{R}^n$ and $\tilde{\mathbf{N}} \in \mathbb{R}^n$ were already defined in (4.17) and (4.19), respectively. The variable $\mathbf{N}_{FF} = \dot{\mathbf{\Gamma}}_{FF}(t)$ is introduced for the subsequent stability analysis.

4.3.2.2 Stability analysis

Before presenting the stability analysis of proposed control solution 1, let us establish the following lemma which is a slight modification of Lemma 1 presented in [133].

Lemma 1. *An auxiliary function called* $\mathbf{L}(t) \in \mathbb{R}^n$ *is defined as:*

$$\mathbf{L}(t) = \mathbf{r}(\mathbf{N}_d(t) - \mathbf{N}_{FF}(t) - \boldsymbol{\beta}\mathrm{sgn}(\mathbf{e}_2)) \qquad (4.28)$$

If the controller gain β is chosen to satisfy the following inequity:

$$\beta > ||\mathbf{N}_d(t)||_{\mathcal{L}\infty} - ||\mathbf{N}_{FF}(t)||_{\mathcal{L}\infty} + \frac{1}{\Lambda_2}\left(||\dot{\mathbf{N}}_d(t)||_{\mathcal{L}\infty} - ||\dot{\mathbf{N}}_{FF}(t)||_{\mathcal{L}\infty}\right) \quad (4.29)$$

then

$$\int_0^t \mathbf{L}(\tau)d\tau \leq \zeta_b \quad (4.30)$$

In which ζ_b is a positive constant defined as:

$$\zeta_b = \beta||\mathbf{e}_2(0)|| + \mathbf{e}_2(0)^T\left(\mathbf{N}_{FF}(0) - \mathbf{N}_d(0)\right) \quad (4.31)$$

Proof. The integral with respect to time of both sides of (4.27) leads to the following [93]:

$$\int_0^t \mathbf{L}(\tau)d\tau = \int_0^t \mathbf{r}(\tau)(\mathbf{N}_d(\tau) - \mathbf{N}_{FF}(\tau) - \beta\mathrm{sgn}(\mathbf{e}_2(\tau)))d\tau \quad (4.32)$$

By substituting the tracking error $\mathbf{r}$ (4.9) into (4.32), we obtain

$$\begin{aligned}
\int_0^t \mathbf{L}(\tau)d\tau &= \int_0^t \Lambda_2\mathbf{e}_2(\tau)^T(\mathbf{N}_d(\tau) - \mathbf{N}_{FF}(\tau) - \beta\mathrm{sgn}(\mathbf{e}_2(\tau)))d\tau \\
&+ \int_0^t \frac{d\mathbf{e}_2(\tau)^T}{dt}\mathbf{N}_d(\tau)d\tau - \int_0^t \frac{d\mathbf{e}_2(\tau)^T}{d\tau}\mathbf{N}_{FF}(\tau)d\tau \quad (4.33)\\
&- \int_0^t \frac{d\mathbf{e}_2(\tau)^T}{d\tau}\beta\mathrm{sgn}(\mathbf{e}_2(\tau)))d\tau
\end{aligned}$$

By performing the integration by parts of the right-hand side of (4.33), the following is obtained:

$$\begin{aligned}
\int_0^t \mathbf{L}(\tau)d\tau &= \int_0^t \Lambda_2\mathbf{e}_2(\tau)^T(\mathbf{N}_d(\tau) - \mathbf{N}_{FF}(\tau))d\tau \\
&- \int_0^t \frac{d\mathbf{e}_2(\tau)^T}{d\tau}\beta\mathrm{sgn}(\mathbf{e}_2(\tau)))d\tau + \mathbf{e}_2(\tau)^T\mathbf{N}_d|_0^t \quad (4.34)\\
&- \int_0^t \mathbf{e}_2(\tau)\frac{d\mathbf{N}_d(\tau)}{d\tau} - \mathbf{e}_2(\tau)^T\mathbf{N}_{FF}|_0^t + \int_0^t \mathbf{e}_2(\tau)\frac{d\mathbf{N}_{FF}(\tau)}{d\tau}
\end{aligned}$$

By rearranging the terms of (4.34), we obtain:

$$\begin{aligned}
\int_0^t \mathbf{L}(\tau)d\tau &= \int_0^t \Lambda_2\mathbf{e}_2(\tau)^T\left(\mathbf{N}_d(\tau) - \mathbf{N}_{FF}(\tau)\right.\\
&\left.+ \frac{1}{\Lambda_2}\left(\frac{d\mathbf{N}_{FF}(\tau)}{d\tau}\frac{d\mathbf{N}_d(\tau)}{d\tau}\right) - \beta\mathrm{sgn}(\mathbf{e}_2(\tau))\right)d\tau \quad (4.35)\\
&+ \mathbf{e}_2^T(t)(\mathbf{N}_d(t) - \mathbf{N}_{FF}(t)) - \beta||\mathbf{e}_2(t)|| + \beta||\mathbf{e}_2(0)||
\end{aligned}$$

By upper bounding the right-hand side of (4.35), the following is obtained:

$$
\int_0^t \mathbf{L}(\tau)d\tau \leq \int_0^t \Lambda_2 \parallel \mathbf{e}_2(\tau) \parallel \left(\parallel \mathbf{N}_d(\tau) \parallel - \parallel \mathbf{N}_{FF}(\tau) \parallel \right.
$$
$$
+ \frac{1}{\Lambda_2} \left(\left\| \frac{d\mathbf{N}_{FF}(\tau)}{d\tau} \right\| - \left\| \frac{d\mathbf{N}_d(\tau)}{d\tau} \right\| \right) - \boldsymbol{\beta} \right) d\tau \qquad (4.36)
$$
$$
+ \parallel \mathbf{e}_2(t) \parallel (\parallel \mathbf{N}_d(t) \parallel - \parallel \mathbf{N}_{FF}(t) \parallel -\boldsymbol{\beta})
$$
$$
+ \boldsymbol{\beta} \parallel \mathbf{e}_2(0) \parallel + \mathbf{e}_2^T(0)(\mathbf{N}_{FF}(0) - \mathbf{N}_d(0))
$$

We can infer from (4.36) that if $\boldsymbol{\beta}$ is selected according to (4.28), then (4.29) is satisfied. □

Theorem 1. *The tracking error in joint space* $\mathbf{e}_1$ *of a PKM whose dynamics is given by (4.6) under the control law (4.22) converges semi-globally asymptotically to zero while the time goes to infinity if the gain K_s is chosen large enough, and the following design parameters are selected such that:*

$$
\Lambda_1 > \frac{1}{2}, \quad \Lambda_2 > 1,
$$
$$
\boldsymbol{\beta} > ||\mathbf{N}_d(t)||_{\mathcal{L}\infty} - ||\mathbf{N}_{FF}(t)||_{\mathcal{L}\infty} + \frac{1}{\Lambda_2} \left(||\dot{\mathbf{N}}_d(t)||_{\mathcal{L}\infty} - ||\dot{\mathbf{N}}_{FF}(t)||_{\mathcal{L}\infty} \right)
$$

Proof. Let $\mathcal{D} \subset \mathbb{R}^{3n+1}$ a domain containing

$$
\mathbf{y}(t) = [\mathbf{z}^T(t) \quad \sqrt{\mathbf{P}(t)}]^T = \mathbf{0} \qquad (4.37)
$$

Where $\mathbf{P}(t) \in \mathbb{R}^n$ is an auxiliary function introduced as:

$$
\mathbf{P}(t) = \boldsymbol{\beta}||\mathbf{e}_2(0)|| - \mathbf{e}_2(0)^T (\mathbf{N}_{FF}(0) - \mathbf{N}_d(0)) - \int_0^t \mathbf{L}(\tau)d\tau \qquad (4.38)
$$

The time-derivative of (4.38) can be written as follows:

$$
\dot{\mathbf{P}}(t) = -\mathbf{r}^T (\mathbf{N}_d(t) - \mathbf{N}_{FF}(t) - \boldsymbol{\beta}\text{sgn}(\mathbf{e}_2)) = -\mathbf{L}(t) \qquad (4.39)
$$

One can note that $\mathbf{P}(t) \geq 0$, $\forall t \geq 0$ taking into account (4.29) and (4.30) □

Now, let us define the following Lyapunov candidate function $\mathbf{V}(\mathbf{y}, t)$: $\mathcal{D} \times [0, \infty) \to \mathbb{R}$ being a continuously differentiable positive-definite function as:

$$
\mathbf{V}(\mathbf{y}, t) = \frac{1}{2}\mathbf{r}^T\mathbf{M}(\mathbf{q})\mathbf{r} + \mathbf{e}_1^T\mathbf{e}_1 + \frac{1}{2}\mathbf{e}_2^T\mathbf{e}_2 + \mathbf{P} \qquad (4.40)
$$

Eq. (4.5) satisfies the following inequality

$$
\lambda_1||\mathbf{y}||^2 \leq \mathbf{V}(\mathbf{y}, t) \leq \lambda_2(||\mathbf{y}||)||\mathbf{y}||^2 \qquad (4.41)
$$

In which

$$\lambda_1 = \frac{1}{2}\min\{1, \underline{m}\}, \quad \lambda_2(||\mathbf{y}||) = \max\{\frac{1}{2}\overline{m}(||\mathbf{y}||), 1\}$$

Computing the time-derivative of (4.40) leads to:

$$\dot{V}(\mathbf{y}, t) = \frac{1}{2}\mathbf{r}^T\dot{\mathbf{M}}(\mathbf{q})\mathbf{r} + \mathbf{r}^T\mathbf{M}(\mathbf{q})\dot{\mathbf{r}} + 2\mathbf{e}_1\dot{\mathbf{e}}_1 + \mathbf{e}_2^T\dot{\mathbf{e}}_2 + \dot{P} \tag{4.42}$$

Substituting (4.39), (4.26), (4.16) and considering (4.7)–(4.9), expression (4.42) can be re-written as:

$$\dot{V}(\mathbf{y}, t) = \mathbf{r}^T\left(-\frac{1}{2}\dot{\mathbf{M}}(\mathbf{q})\mathbf{r} + \mathbf{N}_d + \tilde{\mathbf{N}} - \mathbf{e}_2(t) - ((\mathbf{K}_s + \mathbf{I})\mathbf{r} + \boldsymbol{\beta}\mathrm{sgn}(\mathbf{e}_2)) - \mathbf{N}_{FF}\right)$$
$$+ 2\mathbf{e}_1(\mathbf{e}_2 - \boldsymbol{\Lambda}_1\mathbf{e}_1) + \mathbf{e}_2^T(\mathbf{r} - \boldsymbol{\Lambda}_2\mathbf{e}_2) - \mathbf{r}^T(\mathbf{N}_d - \mathbf{N}_{FF} - \boldsymbol{\beta}\mathrm{sgn}(\mathbf{e}_2))$$
$$+ \frac{1}{2}\mathbf{r}^T\dot{\mathbf{M}}(\mathbf{q})\mathbf{r}$$

$$\tag{4.43}$$

Since $\mathbf{K}_s$, $\mathbf{I}$, $\boldsymbol{\Lambda}_1$, and $\boldsymbol{\Lambda}_2$, are positive definite diagonal feedback gains matrices, we can simplify $\dot{V}(\mathbf{y}, t)$ as:

$$\dot{V}(\mathbf{y}, t) = \mathbf{r}^T\tilde{\mathbf{N}} - (K_s + 1)\mathbf{r}^T\mathbf{r} - 2\boldsymbol{\Lambda}_1\mathbf{e}_1^T\mathbf{e}_1 - \boldsymbol{\Lambda}_2\mathbf{e}_2^T\mathbf{e}_2 + 2\mathbf{e}_2^T\mathbf{e}_1 \tag{4.44}$$

Where the terms $\mathbf{r}^T\mathbf{r}$, $\mathbf{e}_1^T\mathbf{e}_1$, $\mathbf{e}_2^T\mathbf{e}_2$, and $\mathbf{e}_2^T\mathbf{e}_1$ can be upper bounded as follows:

$$\mathbf{r}^T\mathbf{r} \leq ||\mathbf{r}||^2, \quad \mathbf{e}_1^T\mathbf{e}_1 \leq ||\mathbf{e}_1||^2, \quad \mathbf{e}_2^T\mathbf{e}_2 \leq ||\mathbf{e}_2||^2, \quad \mathbf{e}_2^T\mathbf{e}_1 \leq \frac{1}{2}||\mathbf{e}_1||^2 + \frac{1}{2}||\mathbf{e}_2||^2 \tag{4.45}$$

Substituting the inequalities of (4.45) into (4.44) and rearranging terms, $\dot{V}(\mathbf{y}, t)$ can be upper bounded as:

$$\dot{V}(\mathbf{y}, t) \leq \mathbf{r}^T\tilde{\mathbf{N}} + ||\mathbf{e}_1||^2 + ||\mathbf{e}_2||^2 - (K_s + 1)||\mathbf{r}||^2 + 2\Lambda_1||\mathbf{e}_1||^2 - \Lambda_2||\mathbf{e}_2||^2 \tag{4.46}$$

Considering (4.20), the previous equation can be written as:

$$\dot{V}(\mathbf{y}, t) \leq ||\mathbf{r}||\rho(||\mathbf{z}||)||\mathbf{z}|| + ||\mathbf{e}_1||^2 + ||\mathbf{e}_2||^2 - (K_s + 1)||\mathbf{r}||^2$$
$$+ 2\Lambda_1||\mathbf{e}_1||^2 - \Lambda_2||\mathbf{e}_2||^2 \tag{4.47}$$

Rearranging terms, Eq. (4.47) can be expressed as follows:

$$\dot{V}(\mathbf{y}, t) \leq \lambda_3||\mathbf{z}||^2 - \left(K_s||\mathbf{r}||^2 - \rho(||\mathbf{z}||)||\mathbf{z}||||\mathbf{r}||\right) \tag{4.48}$$

Where $\lambda_3 = \min\{2\Lambda_1 - 1, \Lambda_2 - 1, 1\}$, and $\rho(||\mathbf{z}||) \in \mathbb{R}^n$ is a positive globally invertible non-decreasing function [93]. From λ_3 one can see that Λ_1 and

Λ_2 must be chosen according to the following inequalities:

$$\Lambda_1 > \frac{1}{2}, \quad \Lambda_2 > 1 \tag{4.49}$$

By developing the squares of the last term of (4.48), such expression is re-written as:

$$\dot{V}(\mathbf{y}, t) \leq -\left(\lambda_3 - \frac{\rho^2(||\mathbf{z}||)}{4K_s}\right) = c||\mathbf{z}||^2 \tag{4.50}$$

In which, $c||\mathbf{z}||^2$ is a continuous positive semi-definite function defined on the following region

$$\mathcal{D} = \left\{ \mathbf{y} \leq \mathbb{R}^{3n+1} | \ ||\ \mathbf{y}\ || \leq \rho^{-1}\left(2\sqrt{\lambda_3 K_s}\right) \right\} \tag{4.51}$$

The following subset of $\mathcal{D}$ can be defined by:

$$\mathcal{S} = \left\{ \mathbf{y}(t) \subset \mathcal{D} | c||\mathbf{z}||^2 < \lambda_3 \left(\rho^{-1}\left(2\sqrt{\lambda_3 K_s}\right)\right)^2 \right\} \tag{4.52}$$

According to [16], $c||\mathbf{z}||^2$ is consistently continuous in $\mathcal{D}$. So, one can con-clude that:

$$c||\mathbf{z}||^2 \to 0 \quad \text{as} \quad t \to \infty \quad \forall \mathbf{y}(0) \in \mathcal{S} \tag{4.53}$$

Therefore

$$||\mathbf{e}_1|| \to 0 \quad \text{as} \quad t \to \infty \quad \forall \mathbf{y}(0) \in \mathcal{S} \tag{4.54}$$

At this step, the stability proof of this controller is concluded.

4.4 Control solution 2: A RISE feedforward controller with adaptive feedback gains

4.4.1 Motivation

It is well known that for robotic systems developing tasks where the end-effector is in contact with a stiff environment such as grinding, ma-chining, or assembly tasks, a specific interacting wrench is required to en-sure accuracy. Therefore, force contact control may be required. According to [121], the problem of force control can be described as to derive the ac-tuator forces from generating a specified desired wrench (force or torque) at the end-element of the manipulator, when the manipulator is perform-ing its desired motion. Therefore, to implement a force control-loop with the positioning control, it is necessary to have wrench sensors on the end-element or directly on the actuators. SPIDER4 is an RA-PKM designed for machining applications such as milling or drilling in resin materials. Nevertheless, for hardware limitations, it is impossible to integrate any wrench sensor to SPIDER4 in order to implement a force control-loop. As an alternative of the force control-loop, we propose to modify the already presented control solution 1 in the following form. The fixed gains $\mathbf{K}_s$ and $\mathbf{\Lambda}_2$ of the RISE control part will be modified as adaptive time-varying. The purpose of this action is because when the tracking errors start to increase due to contact forces on the material to be machined, corrective action will reduce such tracking error values. If tracking errors become large, then gains $\mathbf{K}_s$ and $\mathbf{\Lambda}_2$ also will increase their values. Nevertheless, if the tracking errors will be at some fair value, the values for $\mathbf{K}_s$ and $\mathbf{\Lambda}_2$ will decrease, adjusting themselves to the best optimum value. The purpose of this modification of the gains is to improve the tracking performance when the robot is performing a machining task and a Pick-and-place operation. Figs. 4.3 and 4.4 illustrate the block diagram of this control scheme applied to Delta PKM and SPIDER4 RA-PKM, respectively. With the above in mind, let us start with the description of control solution 2.

4.4.2 Proposed control law

The proposed control law can be expressed as follows:

$$\mathbf{\Gamma}(t) = \mathbf{\Gamma}_{ARISE} + \mathbf{\Gamma}_{FF} \tag{4.55}$$

Where $\mathbf{\Gamma}_{FF} \in \mathbb{R}^n$ corresponds to the nominal feedforward term described in (4.23), and $\mathbf{\Gamma}_{ARISE} \in \mathbb{R}^n$ is the RISE controller with adaptive feedback gains whose mathematical expression is given as follows:

$$
\begin{aligned}
\mathbf{\Gamma}_{ARISE}(t) = {} & (\mathbf{K}_s(t) + \mathbf{I})\mathbf{e}_2(t) - (\mathbf{K}_s(t) + \mathbf{I})\mathbf{e}_2(0) \\
& + \int_0^t [(\mathbf{K}_s(\tau) + \mathbf{I})\mathbf{\Lambda}_2(\tau)\mathbf{e}_2(\tau) + \beta\mathrm{sgn}(\mathbf{e}_2(\tau))]d\tau
\end{aligned}
\tag{4.56}
$$

Where the adaptive gains $\mathbf{K}_s(t)$ and $\mathbf{\Lambda}_2(t)$ are adjusted by a modification concept of the adaptive gains presented in [99] as follows:

$$\mathbf{K}_s(t) = \bar{\mathbf{K}}_s|\boldsymbol{\eta}| + \mathbf{K}_2 \tag{4.57}$$

$$\mathbf{\Lambda}_2(t) = \bar{\mathbf{\Lambda}}_2|\boldsymbol{\eta}| + \mathbf{K}_3 \tag{4.58}$$

In which, $\bar{\mathbf{K}}_s$ and $\bar{\mathbf{\Lambda}}_2 \in \mathbb{R}^{n \times n}$ are positive-definite constant diagonal matrices used in the adaptation process of the controller feedback gains, and $\mathbf{K}_2, \mathbf{K}_3 \in \mathbb{R}^{n \times n}$ are other positive-definite constant diagonal matrices that establish the minimum possible value for each adaptive feedback gains. $\boldsymbol{\eta} \in \mathbb{R}^n$ is a nonlinear function depending on the combined joint tracking error $\mathbf{e}_2$.

$$\dot{\boldsymbol{\eta}} = \tanh(\mathbf{e}_2) - \boldsymbol{\eta} \tag{4.59}$$

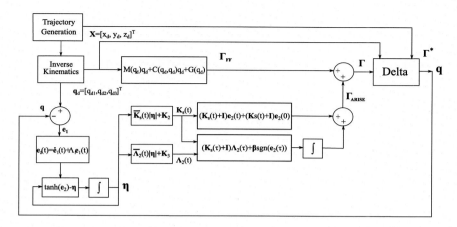

FIGURE 4.3   Block-diagram of proposed control solution 2 for Delta PKM.

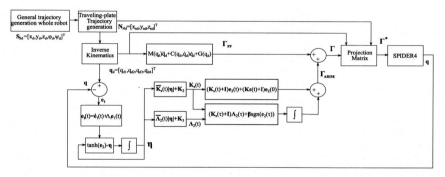

FIGURE 4.4   Block-diagram of proposed control solution 2 for the parallel device of SPIDER4 RA-PKM.

4.4.2.1 *Controller design*

Let us consider the dynamic model of a PKM presented in (4.6) together with the equations defining control solution 2 (4.56)-(4.59). To express the closed-loop control system, it is necessary to reformulate the unmeasurable combined tracking error defined in (4.9) as:

$$\mathbf{r}(t) = \dot{\mathbf{e}}_2(t) + \mathbf{\Lambda}_2(t)\mathbf{e}_2(t) \tag{4.60}$$

Where $\mathbf{\Lambda}_2(t)$ was already defined in (4.58). Multiplying both sides of (4.60) and making use of (4.6) and (4.8), the following is obtained:

$$\begin{aligned}
\mathbf{M}(\mathbf{q})\mathbf{r} = {} & \mathbf{C}(\mathbf{q},\dot{\mathbf{q}})\dot{\mathbf{q}} + \mathbf{G}(\mathbf{q}) + \mathbf{f}(\mathbf{q},\dot{\mathbf{q}}) + \mathbf{\Gamma}_d(t) \\
& + \mathbf{M}(\mathbf{q})(\ddot{\mathbf{q}}_d + \mathbf{\Lambda}_1\dot{\mathbf{e}}_1 + \mathbf{\Lambda}_2(t)\mathbf{e}_2(t)) - \mathbf{\Gamma}_{ARISE}(t) - \mathbf{\Gamma}_{FF}(t)
\end{aligned} \tag{4.61}$$

For stability analysis purposes one can compute the derivative with respect to time of (4.61), yielding the following expression:

$$
\begin{aligned}
\mathbf{M}(\mathbf{q})\dot{\mathbf{r}} = &- \dot{\mathbf{M}}(\mathbf{q})\mathbf{r} + \mathbf{C}(\mathbf{q},\dot{\mathbf{q}})\ddot{\mathbf{q}} + \dot{\mathbf{C}}(\mathbf{q},\dot{\mathbf{q}})\dot{\mathbf{q}} + \dot{\mathbf{G}}(\mathbf{q}) + \dot{\mathbf{f}}(\mathbf{q},\dot{\mathbf{q}}) \\
&+ \dot{\mathbf{\Gamma}}_d(t) + \mathbf{M}(\mathbf{q})\dddot{q}_d + \dot{\mathbf{M}}(\mathbf{q})\ddot{q}_d + \mathbf{\Lambda}_1(\mathbf{M}(\mathbf{q})\ddot{\mathbf{e}}_1 + \dot{\mathbf{M}}(\mathbf{q})\mathbf{e}_1) \\
&+ \mathbf{M}(\mathbf{q})(\mathbf{\Lambda}_2(t)\dot{\mathbf{e}}_2(t) + \dot{\mathbf{\Lambda}}_2(t)\mathbf{e}_2(t)) \\
&+ \dot{\mathbf{M}}(\mathbf{q})\mathbf{\Lambda}_2(t)\mathbf{e}_2(t) - \dot{\mathbf{\Gamma}}_{ARISE} - \dot{\mathbf{\Gamma}}_{FF}
\end{aligned}
\tag{4.62}
$$

Where the time-derivative of $\mathbf{\Gamma}_{ARISE}$ is expressed as:

$$
\dot{\mathbf{\Gamma}}_{ARISE} = (\mathbf{K}_s(t) + \mathbf{I})\dot{\mathbf{e}}_2(t) + [(\mathbf{K}_s(t) + \mathbf{I})\mathbf{\Lambda}_2(t) + \dot{\mathbf{K}}_s(t)]\mathbf{e}_2(t) + \beta\mathrm{sgn}(\mathbf{e}_2(t))
\tag{4.63}
$$

The closed-loop error system can be expressed in the following form:

$$
\mathbf{M}(\mathbf{q})\dot{\mathbf{r}} = -\frac{1}{2}\dot{\mathbf{M}}(\mathbf{q})\mathbf{r} + \mathbf{N}(\mathbf{e}_1, \mathbf{e}_2, \mathbf{r}, t) - \mathbf{e}_2(t) - \dot{\mathbf{\Gamma}}_{ARISE} - \dot{\mathbf{\Gamma}}_{FF}
\tag{4.64}
$$

Where the nonlinear term $\mathbf{N}(\mathbf{e}_1, \mathbf{e}_2, \mathbf{r}, t) \in \mathbb{R}^n$ is written as:

$$
\begin{aligned}
\mathbf{N}(\mathbf{e}_1, \mathbf{e}_2, \mathbf{r}, t) = &\mathbf{C}(\mathbf{q},\dot{\mathbf{q}})\ddot{\mathbf{q}} + \dot{\mathbf{C}}(\mathbf{q},\dot{\mathbf{q}})\dot{\mathbf{q}} + \dot{\mathbf{G}}(\mathbf{q}) + \dot{\mathbf{f}}(\mathbf{q},\dot{\mathbf{q}}) + \dot{\mathbf{\Gamma}}_d(t) + \mathbf{M}(\mathbf{q})\dddot{q}_d \\
&+ \dot{\mathbf{M}}(\mathbf{q})\ddot{q}_d + \mathbf{\Lambda}_1(\mathbf{M}(\mathbf{q})\ddot{\mathbf{e}}_1 + \dot{\mathbf{M}}(\mathbf{q})\mathbf{e}_1) + \mathbf{M}(\mathbf{q})(\mathbf{\Lambda}_2(t)\dot{\mathbf{e}}_2(t) \\
&+ \dot{\mathbf{\Lambda}}_2(t)\mathbf{e}_2(t)) + \dot{\mathbf{M}}(\mathbf{q})\mathbf{\Lambda}_2(t)\mathbf{e}_2(t) + \mathbf{e}_2(t) - \frac{1}{2}\dot{\mathbf{M}}(\mathbf{q})\mathbf{r}
\end{aligned}
\tag{4.65}
$$

For stability analysis purposes, (4.64) is rewritten following the same procedure as in control solution 1.

$$
\mathbf{M}(\mathbf{q})\dot{\mathbf{r}} = -\frac{1}{2}\dot{\mathbf{M}}(\mathbf{q})\mathbf{r} + \mathbf{N}_d + \tilde{\mathbf{N}} - \mathbf{e}_2(t) - \dot{\mathbf{\Gamma}}_{ARISE} - \mathbf{N}_{FF}
\tag{4.66}
$$

In which, $\tilde{\mathbf{N}} = \mathbf{N} - \mathbf{N}_d$ begin $\mathbf{N}_d$ defined in (4.17). Since the nominal feedforward term for this control solution is the same as the previous controller, one can establish $\mathbf{N}_{FF} = \dot{\mathbf{\Gamma}}_{FF}$.

4.4.2.2 Criterion for adaptive gains

Since the nonlinear functions $\mathbf{K}_s(t)$ and $\mathbf{\Lambda}_2(t)$ are continuously differentiable, they can be bounded in the following form:

$$
\mathbf{K}_2 \leq \mathbf{K}_s(t) \leq \mathbf{K}_{sM}
\tag{4.67}
$$

$$
\mathbf{K}_3 \leq \mathbf{\Lambda}_2(t) \leq \mathbf{\Lambda}_{2M}
\tag{4.68}
$$

where $\mathbf{K}_{sM}$ and $\mathbf{\Lambda}_{2M} \in \mathbb{R}^{n \times n}$ represent the maximum admissible values that $\mathbf{K}_s(t)$ and $\mathbf{\Lambda}_2(t)$ may acquire. Moreover, it is assumed that $\mathbf{K}_{sM}$ and

Λ_{2M} exist but are unknown. K_2 and $K_3 \in \mathbb{R}^{n \times n}$ are positive-definite diagonal matrices whose elements denote the minimum possible values that $K_s(t)$ and $\Lambda_2(t)$ can achieve. The values of K_2 and K_3 are determined in the tuning procedure of the controller.

4.4.2.3 Stability analysis

Theorem 2. *The tracking error in joint space e_1 of a PKM whose dynamics is given by (4.6) under the control law (4.55) converges semi-globally asymptotically to zero while the time goes to infinity if the following design parameters are selected such that:*

i) $\Lambda_1 > \dfrac{1}{2}$

ii) $\Lambda_{2M} > K_3$

iii) $\beta > ||N_d(t)||_{\mathcal{L}\infty} - ||N_{FF}(t)||_{\mathcal{L}\infty} + \dfrac{1}{\Lambda_{2M}} \left(||\dot{N}_d(t)||_{\mathcal{L}\infty} - ||\dot{N}_{FF}(t)||_{\mathcal{L}\infty} \right)$

Proof. Considering the same auxiliary function $L(t)$ defined in (4.27), the following inequality is holding:

$$\int_0^t L(\tau)d\tau \leq \beta ||e_2(0)|| + e_2(0)^T (N_{FF}(0) - N_d(0)) \tag{4.69}$$

If β is now selected to satisfy the following condition:

$$\beta > ||N_d(t)||_{\mathcal{L}\infty} - ||N_{FF}(t)||_{\mathcal{L}\infty} + \frac{1}{\Lambda_{2M}} \left(||\dot{N}_d(t)||_{\mathcal{L}\infty} - ||\dot{N}_{FF}(t)||_{\mathcal{L}\infty} \right) \tag{4.70}$$

$\square$

Let us consider the Lyapunov function defined in (4.40)

$$V(y, t) = \frac{1}{2} r^T M(q) r + e_1^T e_1 + \frac{1}{2} e_2^T e_2 + P$$

Developing the time-derivative of (4.40) and making use of (4.39), (4.64), (4.63), (4.60), and (4.8), the following equation results:

$$\dot{V}(y, t) = r^T \tilde{N} - (K_s(t) + 1)r^T r - \dot{K}_s(t) r^T e_2 + 2e_1^T e_2 - 2\Lambda_1 e_1^T e_1 - \Lambda_2(t) e_2^T e_2 \tag{4.71}$$

Considering the lower bounds for $K_s(t)$ and $\Lambda_2(t)$ and the relationships presented in (4.45), expression (4.71) can be bounded as follows:

$$\dot{V}(y, t) \leq ||r|| \rho(|| z ||) ||z|| - (K_2 + 1) ||r||^2 - \frac{|K_{sdm}|}{2} ||r||^2 - \frac{|K_{sdm}|}{2} ||e_2||^2$$
$$+ ||e_1||^2 + ||e_2||^2 - 2\Lambda_1 ||e_1||^2 - K_3 ||e_2||^2 \tag{4.72}$$

Where K_{sdm} denotes a lower bound for $\dot{K}_s(t)$ as is explained in [106]. The previous equation can be re-written in the following form:

$$\dot{V}(\mathbf{y}, t) \leq -\lambda_3||\mathbf{z}||^2 - \left(K_2||\mathbf{r}||^2 - \rho(||\mathbf{z}||)\,||\mathbf{r}||\,||\mathbf{z}|| \right) \tag{4.73}$$

In which $\mathbf{z}(t) \in \mathbb{R}^{3n}$ represents the vector containing the different tracking errors of the system described by Eq. (4.21) and $\lambda_3 = \min\{\eta_1, \eta_2, \eta_3\}$, where the constants η_1, η_2, and η_3 are established as:

$$\eta_1 = 2\Lambda_1 - 1, \quad \eta_2 = \frac{|K_{sdm}|}{2} + K_3 - 1, \quad \eta_3 = \frac{|K_{sdm}|}{2} + 1 \tag{4.74}$$

From (4.74) one can see that Λ_1 must be chosen such that $\Lambda_1 > 1/2$. Completing the squares for the second and last term of (4.73), the following expression is obtained:

$$\dot{V}(\mathbf{y}, t) \leq \lambda_3||\mathbf{z}||^2 - \frac{\rho^2(\mathbf{z})||\mathbf{z}||^2}{4K_2} = -c||\mathbf{z}||^2 \tag{4.75}$$

In (4.75) the term $c||\mathbf{z}||^2$ denotes a continuous positive semi-definite function which is specified in the following domain:

$$\mathcal{D} = \left\{ \mathbf{y} \leq \mathbb{R}^{3n+1} |\, ||\,\mathbf{y}\,|| \leq \rho^{-1}\left(2\sqrt{\lambda_3 K_2} \right) \right\} \tag{4.76}$$

The following subset of $\mathcal{D}$ can be defined by:

$$\mathcal{S} = \left\{ \mathbf{y}(t) \subset \mathcal{D}|c||\mathbf{z}||^2 < \lambda_3 \left(\rho^{-1}\left(2\sqrt{\lambda_3 K_2} \right) \right)^2 \right\} \tag{4.77}$$

According to [16] $c||\mathbf{z}||^2$ is consistently continuous in $\mathcal{D}$. So, one can conclude that:

$$c||\mathbf{z}||^2 \to 0 \quad \text{as} \quad t \to \infty \quad \forall \mathbf{y}(0) \in \mathcal{S} \tag{4.78}$$

Therefore

$$||\mathbf{e}_1|| \to 0 \quad \text{as} \quad t \to \infty \quad \forall \mathbf{y}(0) \in \mathcal{S} \tag{4.79}$$

This concludes the stability proof of the proposed second control solution.

4.5 Conclusion

In this chapter, the contributions related to control for PKMs were presented. Two control solutions based on the RISE control have been developed and analyzed; these control solutions are: A RISE control with nominal feedforward, and a RISE feedforward with adaptive feedback gains.

The first one was proposed to validate the effectiveness of the developed kinematic and dynamic models for Delta PKM and SPIDER4 RA-PKM. The second one was designed to improve the tracking performance of a PKM as SPIDER4 that develops tasks where contact forces are involved such as machining; there are not any feedback sensors to measure such contact forces. However, the second control scheme can also be implemented in PKMs designed for Pick-and-pace applications as the Delta robot. The stability analysis of the proposed controllers was presented in this chapter. In next chapter, we will present simulation and real-time experiments of the proposed control solutions on the experimental platforms described in Chapter 3.

Simulation and real-time experimental results

5.1 Introduction

A key aspect in the design of control schemes is the demonstration of their effectiveness against other control schemes that have been previously proposed. The proposed controllers are based on the RISE controller. Therefore, in this chapter, we present the obtained enhancements of our proposed control solutions compared to the standard RISE controller. The proposed RISE controller with nominal feedforward and RISE feedforward controller with adaptive gains are validated under different scenarios via numerical simulations for Delta PKM and real-time experiments for SPIDER4. For both platforms, control solution 1 was designed mainly to prove the veracity of the inverse dynamic model of both manipulators. For Delta PKM, a pick-and-place trajectory is proposed to be executed in two scenarios. In the first scenario, it is intended that the robot follows this trajectory without any payload, while in the second scenario it is intended that the robot translates a mass of 1 kg. All of this, it should be emphasized, is at the simulation level. In the case of SPIDER4, the control solution 1 is evaluated in a nominal scenario, which is free motion without any physical contact with the spindle tip. The obtained performance for the RISE with nominal feedforward is evaluated against the obtained with the standard RISE control; a machining scenario is used to validate the proposed control solution 2. The machining experiments are carried out on blocks of Styrofoam, which is an excellent material to perform the first machining tests of SPIDER4. The machining scenario is performed under three forward speeds (low, medium, and high) to see how the precision is affected when the same task is executed at different speeds.

　　105

5.2 Performance evaluation criteria

In order to quantify the performance of proposed control solutions RISE FF (proposed control solution 1) and RISE FF AG (proposed control solution 2), the Root Mean Square Error (RMSE) formula is used, which enables us to quantify the accuracy obtained by the tested controller in only one numerical data. This performance index consists of saving a set of samples of the tracking error signals, whose values are squaring. The sum of these values is divided by the total number of samples, and the square root is performed. RMSE has been widely used as a performance index for parallel robots as can be seen in [16], [105], and [49]. For the Delta PKM, the following equations define the RMSE in both Cartesian and joint space are established:

$$RMSE_C = \sqrt{\frac{1}{N} \sum_{k=1}^{N} (e_x^2(k) + e_y^2(k) + e_z^2(k))} \tag{5.1}$$

$$RMSE_J = \sqrt{\frac{1}{N} \sum_{k=1}^{N} (e_{11}^2(k) + e_{12}^2(k) + e_{13}^2(k))} \tag{5.2}$$

where e_x, e_y, e_z denote the Cartesian position tracking error of the traveling plate along the x, y, z axes, while e_{11}, e_{12}, e_{13} are the different joint space tracking errors. Moreover, N is the number of samples and k the current sample. The RMSE expressions for SPIDER4 positioning device in Cartesian ($RMSE_C$) and joint space ($RMSE_J$) are given as follows:

$$RMSE_C = \sqrt{\frac{1}{N} \sum_{k=1}^{N} (e_{xn}^2(k) + e_{yn}^2(k) + e_{zn}^2(k))} \tag{5.3}$$

$$RMSE_J = \sqrt{\frac{1}{N} \sum_{k=1}^{N} (e_{11}^2(k) + e_{12}^2(k) + e_{13}^2(k) + e_{14}^2(k))} \tag{5.4}$$

where e_{xn}, e_{yn}, e_{zn} denote the Cartesian position tracking errors of the traveling-plate along the x, y, z axes, while $e_{11}, e_{12}, e_{13}, e_{14}$ are the tracking errors in joint space of the actuators located in the fixed base.

5.3 Tuning gains procedure

For both controllers tested, the fixed feedback gains were selected via trial and error method. In the case of SPIDER4, there are phenomena such as noise, which can be considerably amplified if the gains are not well

TABLE 5.1 Controller gains values.

DELTA		SPIDER4	
RISE/RISE FF	**RISE FF AG**	**RISE/RISE FF**	**RISE FF AG**
$\Lambda_1 = 100$	$\Lambda_1 = 100$	$\Lambda_1 = 110$	$\Lambda_1 = 110$
$\Lambda_2 = 8$	$\bar{\Lambda}_2 = 4500$	$\Lambda_2 = 1$	$\bar{\Lambda}_2 = 3500$
$K_s = 60$	$\bar{K}_s = 4500$	$K_s = 33$	$\bar{K}_s = 3500$
$\beta = 3$	$\beta = 3$	$\beta = 0.5$	$\beta = 0.5$
	$K_2 = 60$		$K_2 = 33$
	$K_3 = 8$		$K_3 = 1$

selected, damaging the system's performance. Unfortunately for real-time experimentation, it is difficult to perform a gain adjustment analytically when there is noise, so the best solution is to do it by trial and error until proper performance is achieved. In the case of simulations, trial and error tuning allows us to understand the impact of each gain which is reflected in the error signal curves. This makes it easy to perform adjustments to achieve optimal system performance.

5.3.1 Tuning gains procedure for control solution 1

For the proposed RISE control with nominal feedforward and the Standard RISE, firstly, we established a value large enough for Λ_1 and K_s, then we set a value of 1 for Λ_2 subsequently, that value is progressively increased until obtaining the minimum possible tracking error. If the tracking error starts to increase, we proceed to reduce the value of Λ_2. The value of the gain β increases the robustness of the controller. However, this value must be set very small at the beginning and gradually increased in order to avoid the phenomenon of chattering. Table 5.1 in the first column are shown the gains values for the Standard RISE and RISE with nominal feedforward.

5.3.2 Tuning gains procedure for control solution 2

The gains adjustment for this case is similar to the previous one. First, we establish a value large enough for Λ_1, which in this case, is equal to that of scenario 1 because it is the same gain. The gains K_2 and K_3 represent the minimum values that the gains $K_s(t)$ and $\Lambda_2(t)$ can take, respectively. The gains $\overline{K}_s$, and $\overline{\Lambda}_2$ are adjusted according to the desired sensibility to changes of the tracking error in joint space in order to increase the output value of $K_s(t)$ and $\Lambda_2(t)$. Finally, the value for β is kept the same as in control solution 1. The gains values for the RISE feedforward controller with adaptive gains are shown in Table 5.1 in the second column.

5.4 Simulation results for Delta PKM

The Delta robot is a revolutionary PKM whose integration into the industry has been due to its advantages over other types of manipulators in terms of potential accuracy and high operational speed. For such advantages mentioned in previous chapters about the Delta robot architecture, the Polytechnic University of Tulancingo is currently developing a classic prototype with 3-DOF. The prototype aims to be an experimental platform for the validation of advanced nonlinear control techniques. However, there are still issues in the electrical system and the mechanical structure, so the prototype is not entirely ready to experiment with complex control schemes. Due to the limitations presented on the prototype, the control solutions were validated only via numerical simulations using the computed kinematic and dynamic parameters of the prototype. With that said, let us start with the description of the experimental setup of the Delta PKM prototype.

5.4.1 Software settings for simulations

To simulate the proposed control schemes, a Simulink block diagram was programmed containing the following elements:

- A block for programming the generation of trajectories.
- The proposed control schemes were programmed in parallel and run in parallel during the simulation process.
- The equations of the dynamic model of the Delta PKM are replicated three times, in order to execute the three control schemes in parallel.
- The sampling time of the simulation was set to 0.4 ms, which is the same sampling time used in the SPIDER4 real-time experiments.

Fig. 5.1 presents the block diagram made in Simulink used for the simulations of the proposed control schemes applied to the Delta PKM.

5.4.2 Description of the evaluation scenarios

The simulations on the Delta robot were carried out in two scenarios. It is intended that both scenarios follow the same trajectory which is generated using fifth-order polynomial interpolation denoted as follows [87]:

$$x_f = x_i + r(t)\Delta x, \qquad \text{for} \quad 0 \le t \le t_f \tag{5.5}$$

$$r(t) = 10\left(\frac{t}{t_f}\right)^3 - 15\left(\frac{t}{t_f}\right)^4 + 6\left(\frac{t}{t_f}\right)^5 \tag{5.6}$$

Where x_i is the initial position and x_f is the final position, both are given in Cartesian space, $r(t)$ is the trajectory function of two points, $\Delta_x = x_f - x_i$, and t_f is the duration of each displacement of the trajectories, in this case,

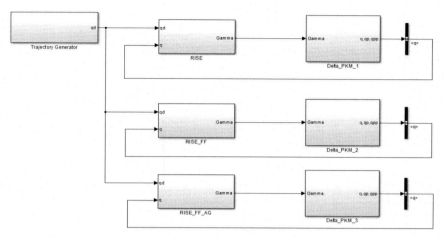

FIGURE 5.1 Block diagram made in Simulink to simulate three control schemes for the Delta PKM.

$t_f = 0.5$ sec. Having detailed the equations that describes the trajectory generation for the Delta PKM, let us describes the evaluation scenarios.

5.4.2.1 Scenario 1 (Nominal scenario)

This scenario consists of trajectory tracking where the Delta PKM moves without carrying any load. Fig. 5.2(a) illustrates the desired trajectory in Cartesian space for this scenario. The generated trajectory is a Pick-and-place kind of path [108], [49], [16]. Although in this scenario it is not intended to move a load, the trajectory used is the same as in the second scenario. We can see in the figure that employing a series of points and the interpolation polynomial, several arc-shaped paths were generated, which try to establish the path that the Delta robot must follow in a pick-and-place operation. The sequence of movements for this Pick-and-place trajectory is presented as follows for the (x, y) plane.

1. Start-Pick: from $(-0.2,-0.1)$ to $(-0.1,0.1)$.
2. Pick-Place: from $(-0.1,0.1)$ to $(0,-0.1)$.
3. Place-Pick: from $(0,-0.1)$ to $(0.1,0.1)$.
4. Pick-Place: from $(0.1,0.1)$ to $(0.2,-0.1)$.
5. Place-Pick: from $(0.2,-0.1)$ to $(0.2,0.1)$.
6. Pick-Place: from $(0.2,0.1)$ to $(-0.2,0.1)$.
7. Place-Pick: from $(-0.2,0.1)$ to $(-0.2,-0.1)$.
8. Pick-Place: from $(-0.2,-0.1)$ to $(0.2,-0.1)$.

The previous movement sequences are performed in 0.3 seconds as can be appreciated in the evolution of the trajectories in Cartesian space depicted in Fig. 5.3.

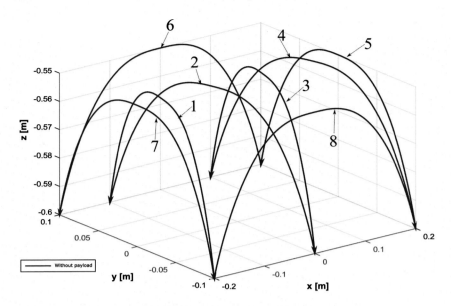

(a) Desired trajectory in Cartesian space for scenario 1

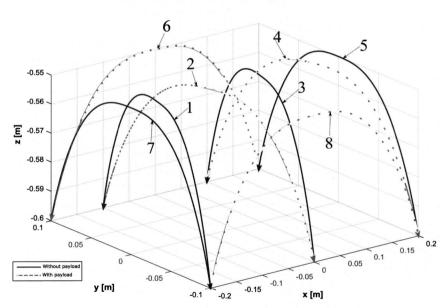

(b) Desired trajectory in Cartesian space for scenario 2

FIGURE 5.2   Desired 3D trajectory for a Pick-and-Place task. The red dotted lines correspond to trajectory segments where the Delta PKM is moving with a payload and the blue solid lines are the corresponding segments without payload.

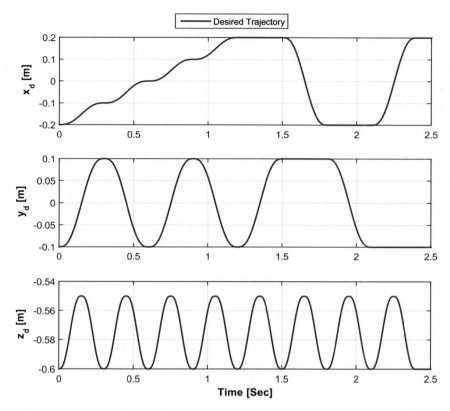

FIGURE 5.3 Evolution of the desired trajectories in Cartesian space versus time for scenarios 1 and 2.

5.4.2.2 Scenario 2 (Pick-and-place scenario)

In this case, we employ the same trajectory described in the previous scenario. However, in certain parts of the path, it is simulated that the robot takes and leaves a load of one kilogram. Such portions of the trajectory in which a mass is moving can be appreciated in Fig. 5.2(b). The sections of the trajectory where the Delta PKM traveling plate moves a mass are depicted with a red dotted line, whereas the solid lines in blue are the trajectory's segment where the Delta PKM is moving without any payload. The sequence of movements for are the same as in the previous case. However, the translation of the payload is described as follows:

1. Start-Pick: from $(-0.2, -0.1)$ to $(-0.1, 0.1)$ without payload.
2. Pick-Place: from $(-0.1, 0.1)$ to $(0, -0.1)$ with payload.
3. Place-Pick: from $(0, -0.1)$ to $(0.1, 0.1)$ without payload.

TABLE 5.2 Controllers performance evaluation for the first scenario (nominal scenario without changes in the payload).

	RMSE$_C$ (cm)	Improvement Cartesian	RMSE$_J$ (deg)	Improvement Joint
RISE	0.0285	0%	0.0490	0%
RISE FF	0.0018	93.70%	0.0045	90.89%
RISE FF AG	0.0015	94.67%	0.0037	92.49%

4. Pick-Place: from (0.1,0.1) to (0.2,−0.1) with payload.
5. Place-Pick: from (0.2,−0.1) to (0.2,0.1) without payload.
6. Pick-Place: from (0.2,0.1) to (−0.2,0.1) with payload.
7. Place-Pick: from (−0.2,0.1) to (−0.2,−0.1) without payload.
8. Pick-Place: from (−0.2,−0.1) to (0.2,−0.1) with payload.

5.4.3 Simulation results of scenario 1

Having described the configuration used to program the control schemes in Simulink as well as the tuning gains and trajectory generation, let us proceed to present the simulation results of the first scenario. As described in Chapter 4, two control schemes were proposed for both the Delta robot and SPIDER4. Simulations were performed considering both control solutions together with the RISE controller in order to carry out a performance evaluation comparison. Fig. 5.4 introduces the evolution of the tracking error of the three control schemes in Cartesian and joint space. The graph clearly shows the superiority of RISE FF and RISE FF AG control schemes over the RISE controller. However, differentiating which of the two proposed control schemes is closer to zero is a bit tricky as they appear to have similar behavior. That is why the performance of the three controllers is quantified using the RMSE formula. These results are registered in Table 5.2, were we can see that both proposed control schemes outperform the RISE controller by more than 90%. However, the percentage improvement of the RISE FF AG adaptive control with respect to the RISE FF is only 1%, so we can conclude at this point that since the system was not subjected to changes in its parameters, the use of an adaptive control scheme does not have a substantial impact on improving the performance of the system. The evolution of the adaptive gains for the RISE FF AG controller is shown in Fig. 5.5. It can be noticed that only at the beginning of the trajectory execution, both gains' sets K_s and Λ_2 have a rapid increase, which decreases over time. Finally, the control signals are shown in Fig. 5.6, where it can be seen that their values do not exceed 8 Nm for the three control schemes.

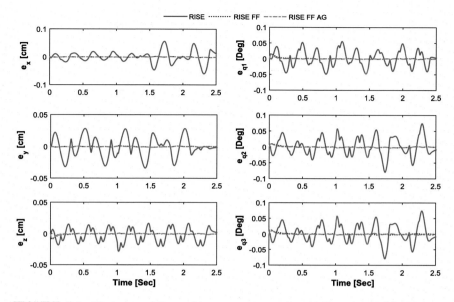

FIGURE 5.4 Evolution of the tracking errors versus time in Cartesian and joint space scenario 1.

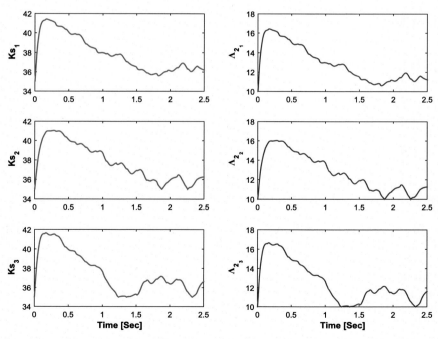

FIGURE 5.5 Evolution of adaptive gains $\mathbf{K}_s$ and Λ_2 scenario 1.

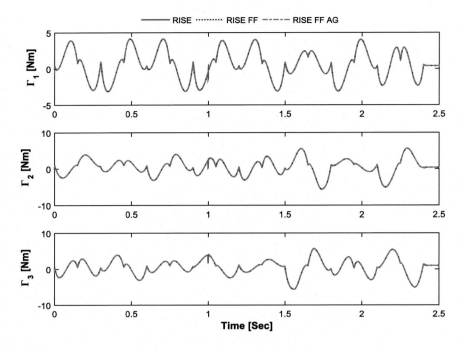

FIGURE 5.6   Evolution of the control inputs of all control schemes scenario 1.

5.4.4 Simulation results of scenario 2

As explained previously, this evaluation scenario incorporates changes in the payload of one kilogram during the execution process. The tracking errors in Cartesian and joint space are presented in Fig. 5.7. This figure shows that the performance of the RISE controller is much lower with respect to the RISE FF and RISE FF AG. It can also be seen in the same figure that the RISE FF AG controller is notably superior to the RISE FF due to the adaptability it offers since the gains are adjusted as the payload value starts to vary. Although the addition of a payload in this scenario affected the performance of the three control schemes, as shown in the RMSEs in Table 5.3, we can see that the RISE FF AG controller outperforms the RISE FF by 14%. We can conclude at this point that the RISE FF AG controller is suitable for pick-and-place tasks since the tracking errors are lower than non-adaptive control schemes. In Fig. 5.8, we can see how the values of the gains K_s and Λ_2 increase in the periods where the robot takes and leaves the payload of one kilogram. Finally, Fig. 5.9 shows the comparison of the control signals, where it can be seen that the values increased considerably compared to the previous scenario (between 5 and 10 Nm more). Having described the results of the implementation of the proposed con-

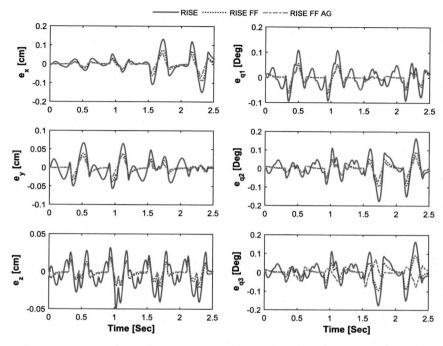

FIGURE 5.7 Evolution of the tracking errors versus time in Cartesian an joint space scenario 2.

TABLE 5.3 Controllers performance evaluation for the second scenario (Pick-and-place).

	RMSE$_C$ (cm)	Improvement Cartesian	RMSE$_J$ (deg)	Improvement Joint
RISE	0.0571	0%	0.0929	0%
RISE FF	0.0312	47.37%	0.0494	46.81%
RISE FF AG	0.0218	61.89%	0.0354	61.90%

trol schemes on the Delta PKM, we will show in the following section the real-time experimental results on SPIDER4 RA-PKM.

5.5 Real-time experimental results for SPIDER4 RA-PKM

5.5.1 Testbed hardware and software description

Fig. 5.10 presents an overview of SPIDER4 [109], [48] in its work cell, including the tooling plate used to fix the material block to be machined. SPIDER4 uses different actuators to perform the linear motions of the

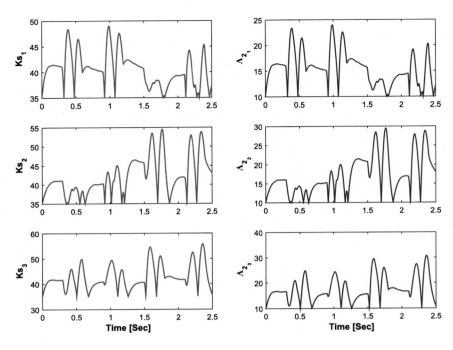

FIGURE 5.8   Evolution of adaptive gains $\mathbf{K}_s$ and $\mathbf{\Lambda}_2$ scenario 2.

parallel positioning device and the wrist. Four WITTESTEIN TPMA110S-022M-6KB1-220H-W6 are responsible for performing the movements on x, y, and z axis. Each actuator includes a gearhead with a gear ratio of 1:22, providing a peak torque of 3100 Nm and 189 rpm as maximum rotation speed. To measure the joint positions, these motors are equipped with multiturn absolute encoders. On the nacelle, three actuators perform the independent angular motions ϕ and ψ as well as the spindle's movement. The motor responsible for the movement in the ϕ axis is STOBER EZH501USVC4P097; this actuator can generate a peak torque of 200 Nm. To perform the angular motion in the ψ axis, a HARMONIC DRIVE CHA-20A-30-H-M1024-B is used. This motor can provide a peak torque of 27 Nm. Finally, the B&R 8JSA24.E4080D000-0 is used for the actuation of the spindle machining tool. This motor can provide a torque of 1.41 Nm and a related speed of 8000 rpm. As was mentioned in Chapter 3, the implemented control solutions affect only the main motors located on the fixed base. However, it is worth mentioning the features of the other motors located on the nacelle. The joint velocities of SPIDER4 are not measured directly because the robot is not equipped with sensors to measure the joint velocities. Notwithstanding, the joint velocities are calculated through numerical derivatives to form the joint positions' measurements with a sam-

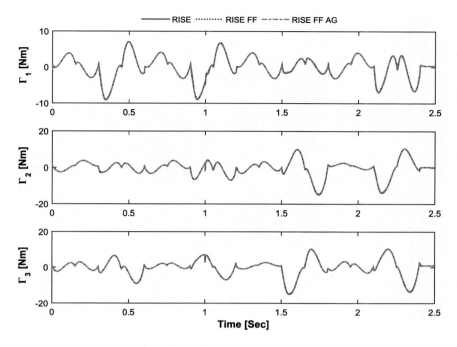

FIGURE 5.9   Evolution of the control inputs of all control schemes scenario 2.

pled time of 0.0004 sec. The structure of each control scheme is created in Simulink software from MathWorks. The Simulink project contains the kinematic algorithms, Jacobian matrices, and the control scheme for SPI-DER4. However, all the functionalities of SPIDER4, including the motion control, are programmed in a B&R Automation Studio project. Therefore, a library called B&R Automation studio target for Simulink is used to convert the Simulink code to C code and transfer it to the B&R Automation Studio Project. When all changes have been carried out to the project, the next step is to compile and transfer them to the B&R Automation PC 910 with a sampling time of 0.4 msec, sending and receiving the control signals to the X20 system which in turn sends and receives signals to the inverted modules called ACOPOS multi system regulating each SPIDER4 motor. The SPIDER4's project is executed through a Graphic User Interface (GUI) developed by Tecnalia Company and programmed in B&R Automation Studio. The experimental setup for SPIDER4 with the main components is illustrated in Fig. 5.11. The evaluation scenarios are two; the first one is the nominal scenario where the robot moves in free motion. The second one is the scenario for machining where the robot follows a desired trajectory with material contact. In this last scenario the robot makes a part on Styrofoam material.

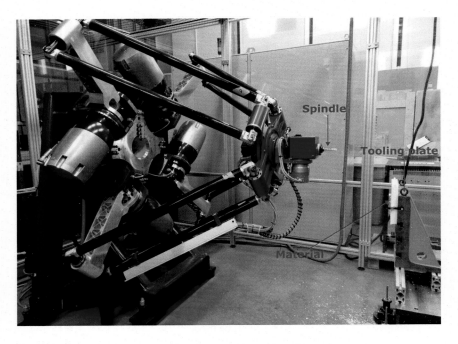

FIGURE 5.10   Overview of SPIDER4 RA-PKM in its work environment.

5.5.2 Reference trajectory generation

To program a machining task on a CNC machine [44], [77], the majority of them used the programming language developed by the Electronics Industry Association (EIA) in the 1960s named RS-274D, better known as "G-code" or "G&M-code", is most often used because many of the instructions of this language starts with the letters G and M. The most common addressed codes described for this programming language are given in Table 5.4. For SPIDER4, the desired trajectories for the tested controllers are generated using the CNC functionality of B&R AUTOMATION STUDIO. There exist many instructions for programming CNC machines using G and M functions; however, only the codes used in the presented experiments on SPIDER4 will be detailed.

- **M3**: Starts the spindle clockwise.
- **M30**: Ends program.
- **F**: Forward speed. For B&R Automation Studio the units are mm/min.
- **G1**: Linear motion. This command moves the tool in a straight line with the programmed forward speed. e.g., Suppose that the spindle of the robot is initially at X0 Y0, and one wants to move it to X10 Y15; the following line code is witted **G1 X10 Y15 F1000** (see Fig. 5.12a).

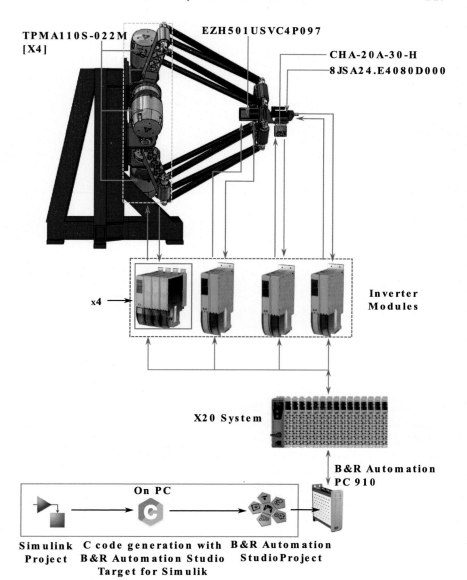

TPMA110S-022M
[X4]

EZH501USVC4P097

CHA-20A-30-H

8JSA24.E4080D000

x4 →

Inverter
Modules

X20 System

B&R Automation
PC 910

On PC

Simulink   C code generation with   B&R Automation
Project   B&R Automation Studio   StudioProject
              Target for Simulik

FIGURE 5.11   Illustration of experimental setup of the SPIDER4.

- **G2**: Circular interpolation clockwise. This command moves the tool in a circle at the programmed forward speed. There are two forms to use this instruction; one is considering the I, J, and K instructions as follows. Suppose that initially, the robot's spindle is in the position X0

TABLE 5.4 Common alphanumeric address codes.

Code	Meaning	Code	Meaning
A	Rotation about X-axis	O	Program number
B	Rotation about Y-axis	P	Dwell time
C	Rotation about Z-axis	Q	Used in drill cycles
F	Forward speed	R	Arc radius
G	G-Code (preparatory code)	S	Spindle speed
H	Tool Length Offset (TLO)	T	Tool number
I	Arc center X-vector	X	X-coordinate
J	Arc center Y-vector	Y	Y-coordinate
K	Arc center Z-vector	Z	Z-coordinate
M	M-Code (miscellaneous code)	N	Block number

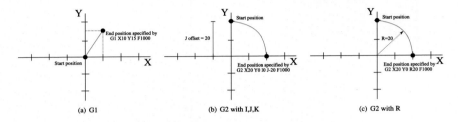

(a) G1   (b) G2 with I,J,K   (c) G2 with R

FIGURE 5.12   Interpolation examples using G1 and G2 instructions.

Y20, and it is desired to move it to point X20 Y0 forming a quarter circle whose center is at position X0 Y0 with a forward speed of 1000 mm/min. The programming line would be expressed as **G2 X20 Y0 I0 J-20 F1000**. Where I and J represent the offset on the X and Y axis, respectively (see Fig. 5.12b for a better understanding). The other way to program G2 is by using the ratio R specification. The previous line code can be written as **G2 X20 Y0 R20 F1000** (see Fig. 5.12c).

- **G3**: Circular interpolation counterclockwise. This command is basically the same as G2, but the interpolation is performed on the opposite side.

5.5.3 Evaluation scenarios

5.5.3.1 *Scenario 1 (Nominal scenario)*

Scenario 1 consists of a free motion trajectory where the SPIDER4 spindle tip is not in contact with any material to be machined. This trajectory is used to validate the RISE control with nominal feedforward. Linear and circular interpolation are used to generate the path to follow. Moreover, the trajectory integrates angular displacements for ϕ and ψ axes whose movements are controlled independently by the proposed controller. The established forward speed F for the entire trajectory is 12000 mm/min. Fig. 5.13 depicts the trajectories' evolution with respect to time and the

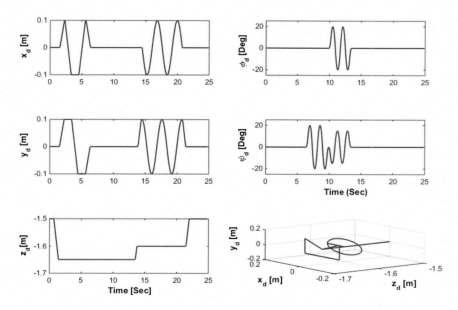

FIGURE 5.13   Evolution of the desired trajectories in Cartesian space versus time.

three-dimensional representation. The points sequence of this trajectory is in Appendix A. However, because the proposed control solutions for SPIDER4 only cover the parallel mechanism without the wrist, it is necessary to provide the reference trajectories just for the nacelle. Nacelle trajectories in Cartesian space are computed evaluating (3.36) with the data provided in Fig. 5.13, and the resulting trajectories for the nacelle are shown in Fig. 5.14.

5.5.3.2 *Scenario 2 (Machining scenario)*

The trajectory generated for this scenario is used for the validation of the RISE feedforward with adaptive gains. Compared to the previous trajectory, this one is more complex due to the large number of operations it contains to generate different geometric shapes. This trajectory describes a milling machining process on a flat piece of Styrofoam with a witness of one inch. The G-code for this milling machining task was firstly simulated in WinUnisoft software from Alecop for a safety implementation on SPIDER4. The cutting depth established for these experiments is 5 mm, where the cutting speed established for the spindle S is 7000 rpm. One can modify the speed S from the G-code, but it is impossible to do it automatically by the established control law. Fig. 5.15 depicts the flat piece to be machined with its dimensions. The trajectory generated by G code is shown in three dimensions in Fig. 5.16, where the lines in blue describe the part

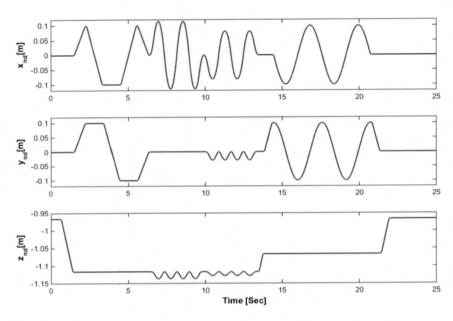

FIGURE 5.14   Desired trajectory of the nacelle in Cartesian space versus time.

of the trajectory where the robot is not cutting material while the lines in red represent the sections of the trajectory where the spindle is devastating the Styrofoam. The Second scenario is executed at different forward speeds to see the impact the speed changes have on the controllers' performance. The speeds values are listed below.

- F=1200 mm/min (low speed)
- F=2400 mm/min (medium speed)
- F=24000 mm/min (high speed)

Fig. 5.17 represents the desired trajectories in Cartesian space for x, y, and z with respect to time executed at the specified forward speeds.

5.6 Experimental results of control solution 1

The experimental results of the RISE controller with nominal feedforward are compared to the standard RISE control in order to see the obtained enhancement when the IDM of the positioning device of SPIDER4 is used as the nominal feedforward term. The different tracking errors in Cartesian and joint space for the Standard RISE and RISE feedforward are depicted in Figs. 5.18 and 5.19, respectively. It can be seen that the proposed RISE controller with nominal feedforward significantly reduces

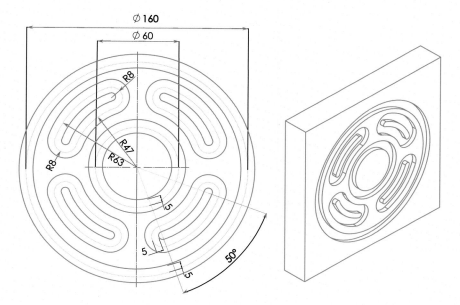

FIGURE 5.15   Scheme of desired piece with dimensions in mm and degrees.

overshooting with respect to the Standard RISE control, even though both tracking errors for the whole trajectory are small. The tracking errors in the Cartesian space depicted in Fig. 5.18 were calculated using the FKM because the controllers are in function of the joint variables since there is no sensor available to measure the nacelle position directly. The generated torques by the RISE feedforward controller are shown in Fig. 5.20, where we can note that the steady-state values for the produced torques Γ_1 and Γ_2 are positive, while the torques Γ_3, and Γ_4 are negative due to the orientation of the kinematic chains and the gravity effects which are different for the upper and the lower chains. From Fig. 5.20, we can also note the presence of noise in the control signals. The noise phenomena came from the fact that measured encoder signals yield noise even if it is not very noticeable. Thereby, when these signals are processed in the control loop, the numerical derivative operations amplify the noise's effect, resulting in noisy control signals. We did not use a filter at the controller's output since it may affect the robustness of the controller; additionally, the computational cost may also increase without significant improvement in the overall system's performance. The computed RMSEs along the desired trajectories for the nacelle presented in Fig. 5.14 are given in Table 5.5. The table also presents the percentage of improvement of the control solution 1 with respect to the Standard RISE controller. As shown in Table 5.5, the

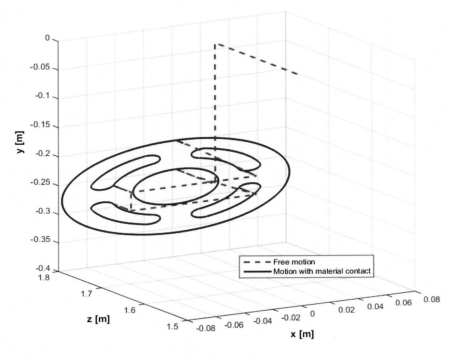

FIGURE 5.16   Desired trajectory for the machining task.

TABLE 5.5   Controllers performance evaluation.

	RMSE$_C$ (cm)	Improvement Cartesian	RMSE$_J$ (deg)	Improvement Joint
RISE	0.0432	0%	0.0552	0%
RISE feedforward	0.0330	23.51%	0.0420	23.95%

RISE feedforward improvement against the standard RISE controller is up to 23%.

5.7 Experimental results of control solution 2

To present the experimental results of control solution 2, Standard RISE and RISE with nominal feedforward controllers are also tested with the same trajectory and execution speeds to view in more detail the impact of the RISE feedforward with adaptive feedback gains in real-time experiments. The three controllers' results are compared to each other in graphs and tables in order to see their effectiveness in a machining task executed at different forward speeds. The name of the proposed controllers is

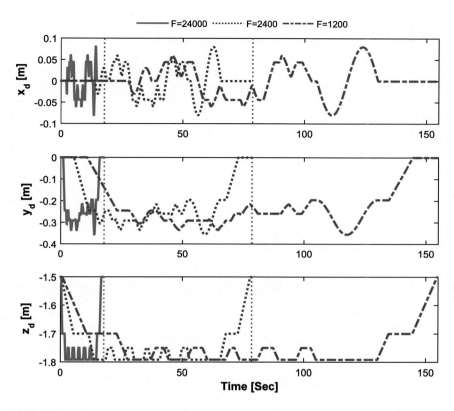

FIGURE 5.17   Evolution of the desired trajectories versus time for different values of F.

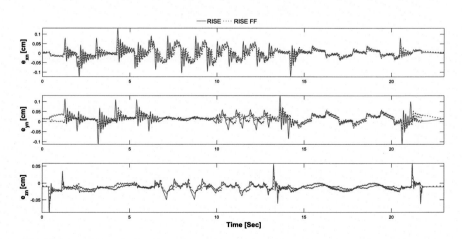

FIGURE 5.18   Evolution of the tracking errors in Cartesian space versus time.

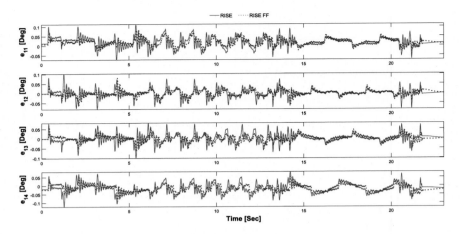

FIGURE 5.19 Evolution of the tracking errors in joint space versus time.

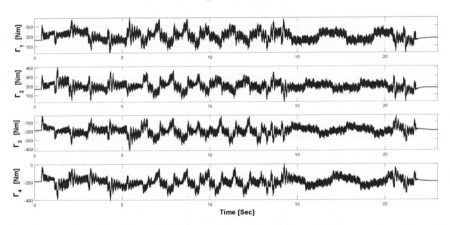

FIGURE 5.20 Evolution of the torques generated by RISE controller with nominal feed-forward versus time.

simplified for presenting the experimental results in graphs and tables as follows: RISE FF represents RISE control with nominal feedforward, and RISE FF AG is RISE feedforward with adaptive gains.

5.7.1 Machining path evaluation at low speed

The comparison of the tracking errors in joint space with a forward speed F = 1200 mm/min is depicted in Fig. 5.21. One can see from the graph for motor 1 that the tracking error in parts is smaller with the RISE controller and in other parts with the RISE FF AG controller. For the graphs of motors 2 and 3, the tracking error is quite similar between the RISE FF

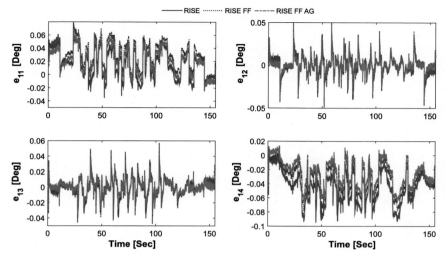

FIGURE 5.21 Evolution of tracking errors in joint space with respect to time for F=1200 mm/min.

and RISE FF AG controllers; however, several overshoots are observed in the standard RISE controller's tracking error. In the graph for motor 4, it can be seen quite well that the tracking error of RISE FF AG is the closest to zero, then RISE FF and finally the standard RISE. The variations in each motors' tracking errors may be due to the fact that although the dynamic parameters of each set of links were considered the same, in reality, there are slight variations in mass between them. Fig. 5.22 is a zoom of the previous figure in the interval of 75 and 80 seconds for a better appreciation of the different curves for F = 1200 mm/min. The evolution of the adaptive gains K_s and Λ_2 are shown in Figs. 5.23 and 5.24, respectively. It can be seen that the minimum values of these gains are those previously established for K_2 and K_3. The maximum peak values that K_s and Λ_2 reach are 36 and 4, respectively. The torques generated from the RISE FF AG controller are illustrated in Fig. 5.25. It can be seen that the values for motors 1 and 2 range between 100 and 30 Nm, while for motors 3 and 4, the values range from −100 to −350 Nm. These variations are due to the horizontal orientation of the SPIDER4 kinematic chains since gravity's acceleration affects the lower linkages (kinematic chains 1 and 2) differently than the upper linkages (kinematic chains 3 and 4). To give a concrete result of the controllers' performance under this operating condition, one must resort to the use of the RMSE equations (5.3) and (5.4); these results are presented in Tables 5.6 and 5.7 for joint and Cartesian space, respectively. With the data provided in these tables, one can compute the resulting improvement of the RISE FF AG with respect to RISE and RISE FF where the results are registered in Table 5.8. We can see that the improvement obtained from the

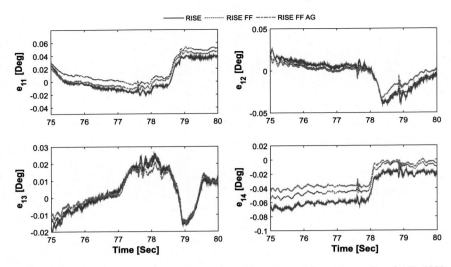

FIGURE 5.22   Evolution of tracking errors in joint space with respect to time for F=1200 mm/min zoomed view.

RISE FF AG controller is 24% in the joint space, and 19% in the Cartesian space compared it to the standard RISE controller. Comparing RISE FF AG with respect to RISE FF, we get improvements of 17% and 15% expressed for joint and Cartesian space. Considering these results, we can note a significant enhancement in the system performance when RISE FF AG is used to regulate the motion of SPIDER4. The resulting machining piece of this experimentation using control solution 2 is presented in Fig. 5.36(a).

5.7.2 Machining path evaluation at medium speed

Increasing the Forward speed F from 1200 to 2400 mm/min, the following results are obtained. In the graphs presented in Fig. 5.26, it can be seen in the four graphs that the tracking error of the RISE FF AG controller is the one that remains closest to zero in most of the execution of the trajectory, then it is the tracking error obtained by the RISE FF. However, the performance of the standard RISE controller was clearly impaired by the increased forward speed. This affirmation can be better seen in Fig. 5.27, where a zoom has been applied between 35 and 40 seconds to appreciate the curves better. Regarding the adaptive gains, it can be seen in Fig. 5.28 for K_s that its maximum peak values increased very little compared to the previous case; it is seen that in some cases it almost reaches values of 37. Moreover, looking at Fig. 5.29, one can note that the behavior of Λ_2 also is like the previous case whose maximum values it reaches reach up to 5. The control signals are presented in Fig. 5.30, where it can be seen that when the forward speed increased, the output torque also increased. We

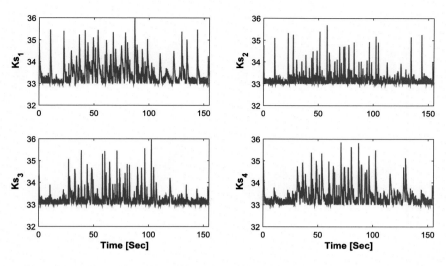

FIGURE 5.23   Evolution of $\mathbf{K}_s$ gains with respect to time for F=1200 mm/min.

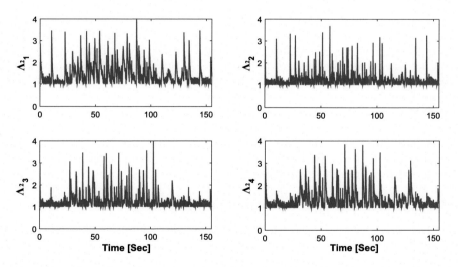

FIGURE 5.24   Evolution of $\mathbf{\Lambda}_2$ gains with respect to time for F=1200 mm/min.

can see that the peak values of motors 1 and 2 are close to 400 Nm, whereas motors 3 and 4 are close to −400 Nm. Using the RMSE formulas, we can confirm the superiority of RISE FF AG in this operative condition concerning the other two controllers. The RMSE results are presented in the third column of Tables 5.6 and 5.7 for Cartesian and joint space, respectively. The obtained improvements with respect to Standard RISE are 21% in the

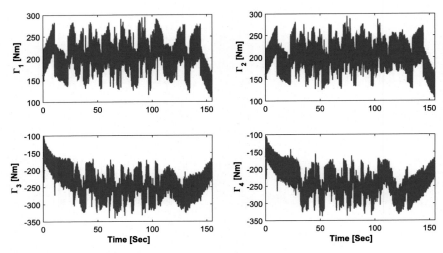

FIGURE 5.25 Evolution of the control inputs generated by RISE FF AG controller with respect to time for F=1200 mm/min.

task space and 25% in joint space. Compared to RISE FF, the acquired enhancements are 16% and 18% for Cartesian and joint spaces. Although the absolute value of the three controllers' tracking errors increased as the forward speed increased, the RISE FF AG controller's tracking error was hardly affected compared to the other two. Hence the percentage of improvements in control solution 2 has increased considerably compared to the previous case. The machining result of the RISE FF AG controller experimentation at this forward speed is shown in Fig. 5.36(b). It is appreciated that the machining quality was low because the spindle was positioned for less time in the material, and its turning speed did not increase as the forward speed increased.

5.7.3 Machining path evaluation at high speed

The forward speed used for this test is not suitable for machining tasks because the quality of the resulting product is quite poor, as we can see in Fig. 5.36(c). Nevertheless, this experiment was proposed to see how the controllers' performance is affected at such a high forward speed. The evolution of the tracking errors in joint space is presented in Fig. 5.31. In comparison to the graphs of the evolution of the tracking errors shown for the previous cases, for this case, it can be seen that the performance of the standard RISE controller is seriously impaired, whereas that the performance of the other two controllers is quite similar. Considering the zoomed view of the described tracking errors in Fig. 5.32, it can be distinguished as a slight improvement of the RISE FF AG over the RISE FF controller. So, it can be intuited that the improvement of the control solution 2 is not much

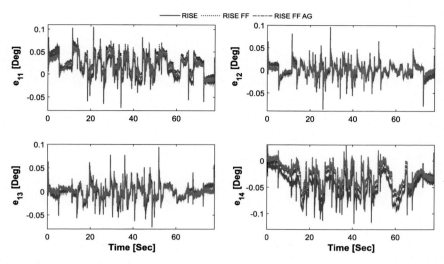

FIGURE 5.26   Evolution of tracking errors in joint space with respect to time for F=2400 mm/min.

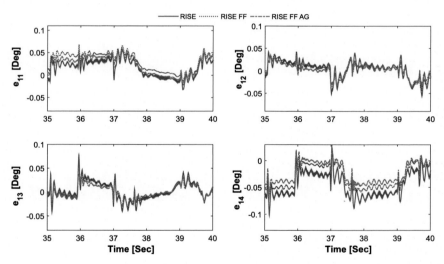

FIGURE 5.27   Evolution of tracking errors in joint space with respect to time for F=2400 mm/min zoomed view.

under this operating condition. Nevertheless, it can be seen in Figs. 5.33 and 5.34 that the maximum values of the adaptive gains increased considerably. One can see that $\mathbf{K}_s$ almost reaches values of 45, and some values of Λ_2 are greater than 10. In Fig. 5.35, the value of the output torques in-

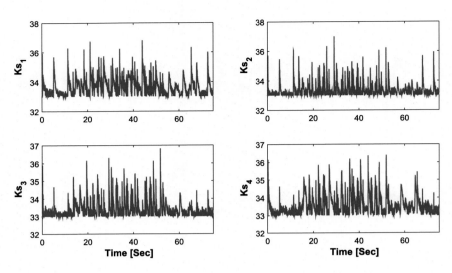

FIGURE 5.28   Evolution of $\mathbf{K}_s$ gains with respect to time for F=2400 mm/min.

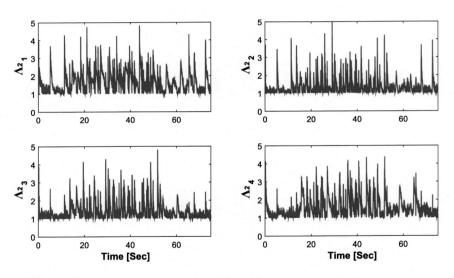

FIGURE 5.29   Evolution of Λ_2 gains with respect to time for F=2400 mm/min.

creased a lot compared to the previous experiments. For motors 1 and 2, these values range from −200 to 600 Nm, and for motors 3 and 4, the peak values range from −600 to 200 Nm. As in previous cases, the RMSEs for this experimentation are registered in Tables 5.6 and 5.7, and the corresponding improvements are in Table 5.8. With F=24000 mm/min, the

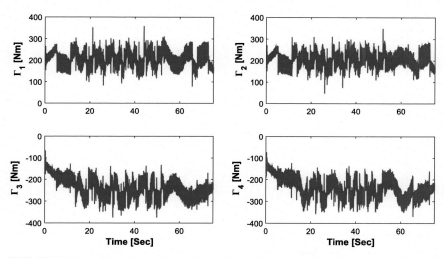

FIGURE 5.30   Evolution of the control inputs generated by RISE FF AG controller with respect to time for F=2400 mm/min.

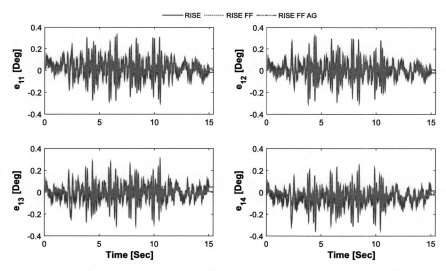

FIGURE 5.31   Evolution of tracking errors in joint space with respect to time for F=24000 mm/min.

improvement with respect to the standard RISE is 45%. However, with respect to the RISE FF controller, the improvement was minimal, only 1.47%, which implies that at high speeds, the performance of the RISE FF AG controller is practically the same as the RISE FF.

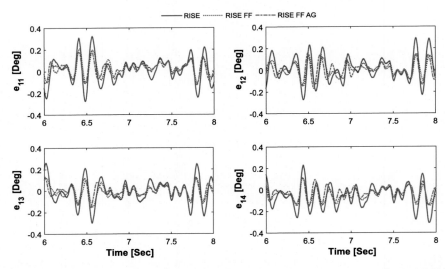

FIGURE 5.32   Evolution of tracking errors in joint space with respect to time for F=24000 mm/min zoomed view.

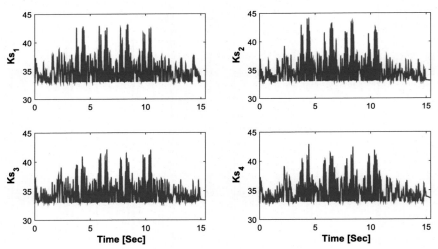

FIGURE 5.33   Evolution of $\mathbf{K}_s$ gains with respect to time for F=24000 mm/min.

TABLE 5.6   Controllers performance evaluation using RMSE for joint space tracking errors (deg).

Controller	Forward speed (F)		
	1200 mm/min	**2400 mm/min**	**24000 mm/min**
RISE	0.0592	0.0618	0.1796
RISE FF	0.0546	0.0566	0.1051
RISE FF AG	0.0449	0.0461	0.1007

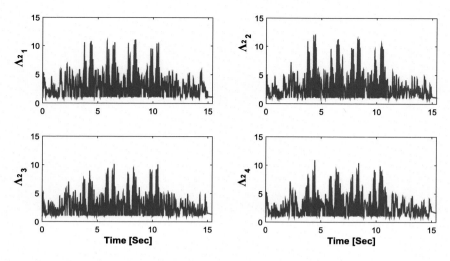

FIGURE 5.34 Evolution of Λ_2 gains with respect to time for F=24000 mm/min.

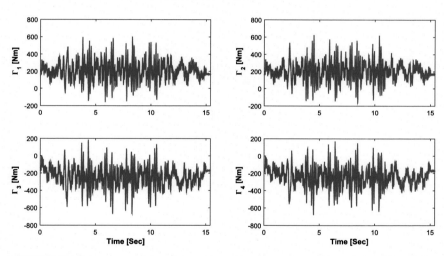

FIGURE 5.35 Evolution of the control inputs generated by RISE FF AG controller with respect to time for F=24000 mm/min.

TABLE 5.7 Controllers performance evaluation using RMSE for Cartesian space tracking errors (cm).

Controller	Forward speed (F)		
	1200 mm/min	2400 mm/min	24000 mm/min
RISE	0.0383	0.0405	0.1425
RISE FF	0.0367	0.0380	0.0794
RISE FF AG	0.0309	0.0319	0.0782

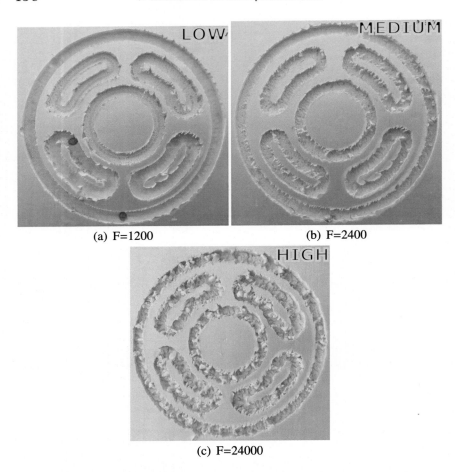

(a) F=1200   (b) F=2400

(c) F=24000

FIGURE 5.36 Machining results using RISE FF AF with different forward speeds.

TABLE 5.8 Enhancement of the proposed RISE FF AG with respect to standard RISE and RISE FF using the RMSE for tracking errors in Cartesian and joint space.

Forward speed	Cartesian RISE	Space RISE FF	Joint RISE	Space RISE FF
Low	19.27%	15.73%	24.155%	17.75%
Medium	21.36%	16.06%	25.40%	18.55%
High	45.12%	1.47%	43.93%	4.18%

5.8 Conclusion

In this chapter, the experimental results related to Delta PKM and SPI-DER4 RA-PKM were presented. In the case of the Delta PKM, the control solutions were validated by numerical simulations whereas for SPIDER4 the proposed controllers were validated by real-time experiments. First, we described the design index to be used for the evaluation of the controllers, in this case, we used the RMSE applied for both the tracking errors in Cartesian space and joint space. Then, the procedure for the tunning of the gains of the tested control schemes was discussed and the values used were detailed employing tables. In the case of the Delta PKM, the way to perform simulations in Simulink was detailed, as well as the algorithm to generate the reference trajectory. This trajectory was used for the evaluation scenarios. The first was the nominal scenario, in which the Delta PKM performed a tracking trajectory without changes in the payload. The second scenario was a pick-and-place task, in which the payload varied by 1 kg in certain portions of the path. The results showed that the RISE FF controller is suitable for the nominal scenario, while the RISE FF AG showed remarkable performance in the pick-and-place task as the adaptive gains kept the tracking error significantly lower than the RISE FF controller and the standard RISE. In the case of SPIDER4, a general overview of the experimental setup was detailed, including a brief description of CNC commands for trajectory generation. Two control solutions were proved on SPIDER4 under two scenarios. The first controller tested on SPIDER4 was the RISE FF. This controller aimed mainly to validate the computed IDM of SPIDER4 since the IDM was used in this controller as the feedforward term. A free-motion trajectory was used in this experiment, and the obtained performance of the control solution 1 was 23% higher than that of the standard RISE controller. The second controller was the RISE FF AG, whose purpose is to compensate for the unknown external disturbances resulting from the friction. Machining is a task where contact forces are involved. Therefore, it is necessary to implement a control scheme to counteract them. The adaptive gains are intended when the error signal increases due to contact with the material, they will increase their value to compensate for this error. By reaching a remarkably small error, the gains will adjust their value optimally. Control solution 2 was validated to control solution 1 and standard RISE in a machining scenario. This scenario was performed with low, medium, and high forward speed. In all speed executions, proposed control solution 2 outperformed the RISE feedforward and standard RISE. However, at high speed, the performance of the RISE feedforward with adaptive gains deteriorates considerably, having a behavior like the RISE feedforward. However, the used maximum speed in these experiments is not suitable for a real machining task. However, the intention of using that speed was to know the limitations of the proposed control schemes precisely. According to the results, we can say that control solution 2 achieves its best performance at medium speed.

General conclusion

The aim of this book has been to present algorithms that allow us to obtain both kinematic and dynamic models of parallel robots with delta architecture and to propose new control strategies to control the positioning of these devices. The verification of the proposed mathematical models (kinematic and dynamic); was performed by testing in simulation in the case of the Delta PKM and real-time experiments in the case of SPIDER4. In the same way, this book showed the effectiveness of the proposed control schemes a RISE controller with nominal feedforward (RISE FF), and RISE feedforward whit adaptive feedback gains (RISE FF AG).

Summary of the work

As mentioned in Chapter 1, the performance of a parallel robot is a function of various factors themselves, such as mechanical design, mathematical modeling, and the control scheme. All these factors must be well designed to get the most out of the closed-loop architecture of PKMs. This book is focused on the modeling and control of PKMs. Control of PKMs involves many challenges, such as highly nonlinear dynamics, the presence of parametric and non-parametric uncertainties, and in some cases, redundant actuation. Therefore, the controllers that must be implemented need to guarantee good precision in the task that the manipulator is developing despite the problems mentioned above. The contributions of the book can be summarized as follows:

1. We presented the kinematic a dynamic modeling formulation of a 3 DOF Delta PKM. Although the algorithms presented are found in the literature, in this book, we presented some MATLAB® codes for programming inverse, direct kinematics, and the Jacobian matrix. These programming code examples are useful for the reader to better understand the context of parallel robots.
2. The kinematic and dynamic models for SPIDER4 RA-PKM. This manipulator is formed by a delta-like parallel positioning device and a serial wrist mechanism that performs the spindle's orientation. The parallel device's Inverse Kinematic Model (IKM) is an adaptation of the 3-DOF Delta PKM, while the IKM of the wrist was obtained by employing transformation matrices as in serial robots. The parallel device's Forward Kinematic Model (FKM) was obtained through a virtual sphere

intercession algorithm, while that of the wrist was calculated using the same transformation matrices used for its IKM. The Inverse Dynamic Model (IDM) of the SPIDER4 parallel positioning device was calculated based on the method developed in [84] for robots with delta-type architecture. The main modification lies in the formulation of the acceleration of gravity in the kinematic chains of SPIDER4 since these are oriented horizontally instead of vertically, which directly affects the motors' output torque. The dynamic model of the wrist was calculated using the Euler-Lagrange formulation. In the end, both models were combined to express the total SPIDER4 IDM in the joint space. However, for the experimental implementation of the proposed control solutions, only the IDM of the parallel positioning device was considered because it was not possible to modify the wrist control law due to software limitations. Finally, the kinematic and dynamic parameters of SPIDER4 RA-PKM were obtained through datasheets and SolidWorks calculations.

3. A RISE controller with nominal feedforward was proposed to validate the IDM of the Delta PKM and the SPIDER4 parallel positioning device through numerical simulations (for Delta) and real-time experiments (for SPIDER4). It is well known that the total or partial integration of the IDM in a control scheme markedly improves the system's performance to be controlled. However, a bad approximation of the model, together with an imperfect estimation of the dynamic parameters, can degrade the system's performance instead of improving it. The proposed RISE feedforward controller was tested together with the standard RISE control in order to see the improvement produced by incorporating the dynamic model proposed as a lead compensation term. The obtained experimental results showed an important improvement with respect to standard RISE control.

4. SPIDER4 RA-PKM is designed to perform machining tasks such as drilling or milling in resin materials. Due to the existence of contact forces, some CNC machines incorporate force controllers to counteract these forces. In order to implement a force controller, the robot must have sensors that measure the contact force or the current consumed by the robot's actuators. However, SPIDER4 does not have this type of sensor to implement a force control loop. So, it was proposed to modify the RISE feedforward controller by replacing fixed feedback gains with adaptive gains. The purpose of these gains is that when the tracking errors grow noticeably due to contact forces, the gains automatically increase their values to counteract such tracking errors. By reducing the tracking error, the gains will automatically adjust to the most optimal value. The validation of this control scheme was carried out through real-time experiments. The experiments consisted of the machining of a piece of Styrofoam. The machining trajectory was

executed with different advanced speeds to see its impact on the SPI-DER4's performance. The RISE feedforward controller's performance with adaptive gains was compared with the obtained using the RISE feedforward and the standard RISE controllers. The obtained results showed a very noticeable improvement with respect to the other controllers when the machining task was executed with a forward speed F=1200 mm/min and F=2400 mm/min. However, when the speed was exceeded at F=24000 mm/min, the performance obtained was slightly higher than that of RISE feedforward. Therefore, it can be concluded that this control proposal is suitable for machining tasks in soft materials if the forward speed is not exceeded. This control scheme was also validated by numerical simulations on the Delta robot showing that it is a good solution for pick-and-place tasks since the controller gains were adjusted to compensate for the tracking errors produced by the variations in the payload.

A

Trajectory points for SPIDER4 real-time experiments

This appendix presents the values of the commands used to generate trajectories in the SPIDER4 workspace for the nominal scenario (scenario 1) and the machining scenario (scenario 2). The units for x, y, and z are given in mm, whereas that for ϕ, ψ, I, K, H, and R are in degrees.

A.1 Trajectory points for scenario 1

The sequence of points shown in Table A.1 below produces the graph of the trajectory for the first scenario; which can be viewed in the third dimension in Fig. 5.13.

TABLE A.1 List of commands used for trajectory generation for scenario 1.

Line	G command	x	y	z	ϕ	ψ	I	J	H	R
N01	G1	0	0	-1500	0	0	0	0	0	0
N02	G1	0	0	-1650	0	0	0	0	0	0
N03	G1	100	100	-1650	0	0	0	0	0	0
N04	G1	-100	100	-1650	0	0	0	0	0	0
N05	G1	-100	-100	-1650	0	0	0	0	0	0
N06	G1	100	-100	-1650	0	0	0	0	0	0
N07	G1	0	0	-1650	0	0	0	0	0	0
N08	G1	0	0	-1650	20	0	0	0	0	0
N09	G1	0	0	-1650	-20	0	0	0	0	0
N10	G1	0	0	-1650	20	0	0	0	0	0
N11	G1	0	0	-1650	-20	0	0	0	0	0
N12	G1	0	0	-1650	0	0	0	0	0	0
N13	G1	0	0	-1650	-15	20	0	0	0	0

continued on next page

143

TABLE A.1 (*continued*)

Line	G command	x	y	z	ϕ	ψ	I	J	H	R
N14	G1	0	0	−1650	15	−20	0	0	0	0
N15	G1	0	0	−1650	−15	20	0	0	0	0
N16	G1	0	0	−1650	15	−20	0	0	0	0
N17	G1	0	0	−1650	0	0	0	0	0	0
N18	G1	0	0	−1600	0	0	0	0	0	0
N19	G1	0	100	−1600	0	0	0	0	0	0
N20	G3	0	100	−1600	0	0	0	−100	720	0
N21	G1	0	100	−1600	0	0	0	0	0	0
N22	G1	0	100	−1500	0	0	0	0	0	0

A.2 Trajectory points for scenario 2

The sequence of points shown in Table A.2 below produces the graph of the trajectory for the second scenario; which can be viewed in the third dimension in Fig. 5.16.

TABLE A.2 List of commands used for trajectory generation for scenario 2.

Line	G command	x	y	z	ϕ	ψ	I	J	H	R
N01	G1	0	0	−1500	0	0	0	0	0	0
N02	G1	0	0	−1700	0	0	0	0	0	0
N03	G1	0	−245	−1700	0	0	0	0	0	0
N04	G1	0	−245	−1792	0	0	0	0	0	0
N05	G3	0	−245	−1792	0	0	0	−30	0	0
N06	G1	0	−245	−1750	0	0	0	0	0	0
N07	G1	44.166	−291.074	−1750	0	0	0	0	0	0
N08	G1	44.166	−291.074	−1792	0	0	0	0	0	0
N09	G2	59.194	−296.548	−1792	0	0	0	0	0	8
N10	G2	21.548	−334.194	−1792	0	0	0	0	0	63
N11	G2	16.074	−319.166	−1792	0	0	0	0	0	8
N12	G3	44.166	−291.074	−1792	0	0	0	0	0	47
N13	G1	44.166	−291.074	−1750	0	0	0	0	0	0
N14	G1	−44.166	−291.074	−1750	0	0	0	0	0	0
N15	G1	−44.166	−291.074	−1792	0	0	0	0	0	0
N16	G3	−59.194	−296.548	−1792	0	0	0	0	0	8
N17	G3	−21.548	−334.194	−1792	0	0	0	0	0	63
N18	G3	−16.074	−319.166	−1792	0	0	0	0	0	8
N19	G2	−44.166	−291.074	−1792	0	0	0	0	0	47
N20	G1	−44.166	−291.074	−1750	0	0	0	0	0	0
N21	G1	−44.166	−258.926	−1750	0	0	0	0	0	0
N22	G1	−44.166	−258.926	−1792	0	0	0	0	0	0
N23	G2	−59.194	−253.452	−1792	0	0	0	0	0	8

continued on next page

TABLE A.2 (*continued*)

Line	G command	x	y	z	ϕ	ψ	I	J	H	R
N24	G2	−21.548	−215.806	−1792	0	0	0	0	0	63
N25	G2	−16.074	−230.834	−1792	0	0	0	0	0	8
N26	G3	−44.166	−258.926	−1792	0	0	0	0	0	47
N27	G1	−44.166	−258.926	−1750	0	0	0	0	0	0
N28	G1	44.166	−258.926	−1750	0	0	0	0	0	0
N29	G1	44.166	−258.926	−1792	0	0	0	0	0	0
N30	G3	59.194	−253.452	−1792	0	0	0	0	0	8
N31	G3	21.548	−215.806	−1792	0	0	0	0	0	63
N32	G3	16.074	−230.834	−1792	0	0	0	0	0	8
N33	G2	44.166	−258.926	−1792	0	0	0	0	0	47
N34	G1	44.166	−258.926	−1750	0	0	0	0	0	0
N35	G1	0	−195	−1750	0	0	0	0	0	0
N36	G1	0	−195	−1792	0	0	0	0	0	0
N37	G3	0	−195	−1792	0	0	0	−80	0	0
N38	G1	0	−195	−1700	0	0	0	0	0	0
N39	G1	0	0	−1700	0	0	0	0	0	0
N40	G1	0	0	−1500	0	0	0	0	0	0

Bibliography

[1] M. Ahmadi, M. Dehghani, M. Eghtesad, A.R. Khayatian, Inverse dynamics of hexa parallel robot using lagrangian dynamics formulation, in: International Conference on Intelligent Engineering Systems, vol. 2008, IEEE, 2008, pp. 145–149.

[2] C.H. An, C.H. Atkeson, J.D. Griffiths, J.M. Hollerbach, Experimental evaluation of feedforward and computed torque control, in: Proceedings. 1987 IEEE International Conference on Robotics and Automation, vol. 4, IEEE, 1987, pp. 165–168.

[3] B. Armstrong, B.A. Wade, Nonlinear PID control with partial state knowledge: Damping without derivatives, The International Journal of Robotics Research 19 (8) (2000) 715–731.

[4] U. Asif, Design of a parallel robot with a large workspace for the functional evaluation of aircraft dynamics beyond the nominal flight envelope, International Journal of Advanced Robotic Systems 9 (2) (2012) 51.

[5] F. Azad, S. Rad, M. Arashpour, Back-stepping control of delta parallel robots with smart dynamic model selection for construction applications, Automation in Construction 137 (2022) 104211.

[6] S. Azizi, R. Soleimani, M. Ahmadi, A. Malekan, L. Abualigah, F. Dashtiahangar, Performance enhancement of an uncertain nonlinear medical robot with optimal nonlinear robust controller, Computers in Biology and Medicine 146 (2022) 105567.

[7] C. Baradat, V. Nabat, S. Krut, F. Pierrot, Par2: A spatial mechanism for fast planar, 2-dof, pick-and-place applications, in: Fundamental Issues and Future Research Directions for Parallel Mechanisms and Manipulators, 2009, p. 10.

[8] M. Bennehar, A. Chemori, F. Pierrot, A new extension of desired compensation adaptive control and its real-time application to redundantly actuated PKMs, in: 2014 IEEE/RSJ International Conference on Intelligent Robots and Systems, IEEE, 2014, pp. 1670–1675.

[9] M. Bennehar, A. Chemori, F. Pierrot, A novel rise-based adaptive feedforward controller for redundantly actuated parallel manipulators, in: 2014 IEEE/RSJ International Conference on Intelligent Robots and Systems, IEEE, 2014, pp. 2389–2394.

[10] M. Bennehar, A. Chemori, F. Pierrot, V. Creuze, Extended model-based feedforward compensation in L1 adaptive control for mechanical manipulators: Design and experiments, Frontiers in Robotics and AI 2 (2015) 32.

[11] M. Bennehar, A. Chemori, F. Pierrot, L1 adaptive control of parallel kinematic manipulators: Design and real-time experiments, in: 2015 IEEE International Conference on Robotics and Automation (ICRA), IEEE, 2015, pp. 1587–1592.

[12] M. Bennehar, Some Contributions to Nonlinear Adaptive Control of PKMs: From Design to Real-Time Experiments, Université de Montpellier II, 2015.

[13] M. Bennehar, A. Chemori, S. Krut, F. Pierrot, Adaptive control of parallel manipulators: Design and real-time experiments, 2016.

[14] M. Bennehar, A. Chemori, F. Pierrot, A new revised desired compensation adaptive control for enhanced tracking: Application to RA-PKMs, Advanced Robotics 30 (17–18) (2016) 1199–1214.

[15] M. Bennehar, G. El-Ghazaly, A. Chemori, F. Pierrot, A novel adaptive terminal sliding mode control for parallel manipulators: Design and real-time experiments, in: 2017 IEEE International Conference on Robotics and Automation (ICRA), IEEE, 2017, pp. 6086–6092.

[16] M. Bennehar, A. Chemori, M. Bouri, L.F. Jenni, F. Pierrot, A new RISE-based adaptive control of PKMs: Design, stability analysis and experiments, International Journal of Control 91 (3) (2018) 593–607.

[17] L. Birglen, Haptic devices based on parallel mechanisms. State of the art, http://www.parallemic.org/Reviews/Review003.html, 2003.

[18] I. Bonev, The true origins of parallel robots, http://www.parallemic.org/Reviews/Review007.html, 2003.

[19] G. Borchert, M. Battistelli, G. Runge, A. Raatz, Analysis of the mass distribution of a functionally extended delta robot, Robotics and Computer-Integrated Manufacturing 31 (2015) 111–120.

[20] S. Briot, C. Baradat, V. Arakelian, Contribution to the mechanical behavior improvement of the robotic navigation device Surgiscope®, in: ASME 2007 International Design Engineering Technical Conferences and Computers and Information in Engineering Conference, American Society of Mechanical Engineers Digital Collection, 2007, pp. 653–661.

[21] S. Briot, W. Khalil, Dynamics of Parallel Robots: From Rigid Bodies to Flexible Elements, vol. 35, Springer, 2015.

[22] S. Briot, V. Rosenzveig, P. Martinet, E. Özgür, N. Bouton, Minimal representation for the control of parallel robots via leg observation considering a hidden robot model, Mechanism and Machine Theory 106 (2016) 115–147.

[23] T. Campbell, C. Williams, O. Ivanova, B. Garrett, Could 3D Printing Change the World? Technologies, Potential, and Implications of Additive Manufacturing, Atlantic Council, Washington, DC, 2011.

[24] C. Cao, N. Hovakimyan, Design and analysis of a novel L1 adaptive controller, part I: Control signal and asymptotic stability, in: 2006 American Control Conference, IEEE, 2006, pp. 3397–3402.

[25] C. Cao, N. Hovakimyan, Design and analysis of a novel L1 adaptive controller, part II: Guaranteed transient performance, in: 2006 American Control Conference, IEEE, 2006, pp. 3403–3408.

[26] G. Carabin, L. Scalera, T. Wongratanaphisan, R. Vidoni, An energy-efficient approach for 3D printing with a Linear Delta Robot equipped with optimal springs, Robotics and Computer-Integrated Manufacturing 67 (2021) 102045.

[27] K.L. Cappel, Motion simulator, Google Patents, US Patent RE27,051, 1971.

[28] L.A. Castañeda, A. Luviano-Juárez, I. Chairez, Robust trajectory tracking of a delta robot through adaptive active disturbance rejection control, IEEE Transactions on Control Systems Technology 23 (4) (2014) 1387–1398.

[29] M. Ceccarelli, E. Ottaviano, Design problems for parallel manipulators in assembling operations, IFAC Proceedings Volumes 36 (23) (2003) 13–26.

[30] D. Chablat, P. Wenger, A new three-DOF parallel mechanism: Milling machine applications, arXiv, 2007.

[31] A. Chemori, Control of complex robotic systems: Challenges, design and experiments, in: 2017 22nd International Conference on Methods and Models in Automation and Robotics (MMAR), IEEE, 2017, pp. 622–631.

[32] H. Cheng, Y.K. Yiu, Z. Li, Dynamics and control of redundantly actuated parallel manipulators, IEEE/ASME Transactions on Mechatronics 8 (4) (2003) 483–491.

[33] X. Chen, X.J. Liu, F. Xie, T. Sun, A comparison study on motion/force transmissibility of two typical 3-DOF parallel manipulators: The sprint Z3 and A3 tool heads, International Journal of Advanced Robotic Systems 11 (1) (2014) 5.

[34] X. Chen, X. Liang, X. Sun, Y. Deng, W. Feng, Y. Cao, Workspace and statics analysis of 4-UPS-UPU parallel coordinate measuring machine, Measurement 55 (2014) 402–407.

[35] L. Cheng, D. Li, G. Yu, Z. Zhang, S. Yu, Robotic arm control system based on brain-muscle mixed signals, Biomedical Signal Processing and Control 77 (2022) 103754.

[36] R. Clavel, Delta, a fast robot with parallel geometry, in: Proc. 18th Int. Symp. on Industrial Robots, Lausanne, 1988, pp. 91–100.

[37] R. Clavel, Device for the movement and positioning of an element in space, Google Patents, US Patent 4,976,582, 1990.

[38] Reymond Clavel, Conception d'un robot parallèle rapide à 4 degrés de liberté, EPFL, 1991.

[39] Alain Codourey, Contribution à la commande des robots rapides et précis, EPFL, 1991.

[40] A. Codourey, Dynamic modelling and mass matrix evaluation of the DELTA parallel robot for axes decoupling control, in: Proceedings of IEEE/RSJ International Conference on Intelligent Robots and Systems, IROS'96, vol. 3, IEEE, 1996, pp. 1211–1218.

[41] A. Codourey, Dynamic modeling of parallel robots for computed-torque control implementation, The International Journal of Robotics Research 17 (12) (1998) 1325–1336.

[42] O. Company, F. Pierrot, A new 3T-1R parallel robot, in: Proc. of IEEE ICAR'99: 9th International Conference on Advanced Robotics, Tokyo, Japan, 1999, pp. 557–562.

[43] D. Corbel, M. Gouttefarde, O. Company, F. Pierrot, Towards 100G with PKM. Is actuation redundancy a good solution for pick-and-place?, in: 2010 IEEE International Conference on Robotics and Automation, IEEE, 2010, pp. 4675–4682.

[44] M.F. Corapsiz, K. Erenturk, Trajectory tracking control and contouring performance of three-dimensional CNC, IEEE Transactions on Industrial Electronics 63 (4) (2016) 2212–2220.

[45] J.J. Craig, P. Hsu, S.S. Sastry, Adaptive control of mechanical manipulators, The International Journal of Robotics Research 6 (2) (1987) 16–28.

[46] EcoRobotix, ARA switch to smart scouting, https://www.ecorobotix.com/en/autonomous-scouting-robot/, 2011.

[47] J.M. Escorcia-Hernandez, H. Aguilar-Sierra, A. Chemori, H. Arroyo-Nunez, O. Aguilar-Meja, An intelligent compensation trough B-spline neural network for a Delta parallel robot, in: Proc. 6th International Conference on Control, Decision and Information Technologies, Paris, France, 2019.

[48] J.M. Escorcia-Hernandez, A. Chemori, H. Aguilar-Sierra, J.A. Monroy-Anieva, A new solution for machining with RA-PKMs: Modelling, control and experiments, Mechanism and Machine Theory 150 (2020) 103864.

[49] J.M. Escorcia-Hernandez, H. Aguilar-Sierra, O. Aguilar-Meja, A. Chemori, H. Arroyo-Nunez, A new adaptive RISE feedforward approach based on associative memory neural networks for the control of PKMs, Journal of Intelligent & Robotic Systems 100 (2020) 827–847.

[50] S. Fan, S. Fan, W. Lan, G. Song, A new approach to enhance the stiffness of heavy-load parallel robots by means of the component selection, Robotics and Computer-Integrated Manufacturing 61 (2020) 101834.

[51] A. Fay, J. Perrin, E. Féry-Lemonnier, Classification des systèmes de robotique chirurgicale et de chirurgie assistée par ordinateur, ITBM-RBM 23 (6) (2002) 326–332.

[52] Y. Feng, X. Yu, Z. Man, Non-singular terminal sliding mode control of rigid manipulators, Automatica 38 (12) (2002) 2159–2167.

[53] ForceDimension, sigma.7 haptic device, https://www.forcedimension.com/products/sigma-7/overview, 2010.

[54] Z. Gao, Active disturbance rejection control: A paradigm shift in feedback control system design, in: 2006 American Control Conference, IEEE, 2006, 7 pp.

[55] Z. Geng, L.S. Haynes, J.D. Lee, R.L. Carroll, On the dynamic model and kinematic analysis of a class of Stewart platforms, Robotics and Autonomous Systems 9 (4) (1992) 237–254.

[56] C. Germain, S. Briot, V. Glazunov, S. Caro, P. Wenger, Irsbot-2: A novel two-dof parallel robot for high-speed operations, in: ASME 2011 International Design Engineering Technical Conferences and Computers and Information in Engineering Conference, American Society of Mechanical Engineers Digital Collection, 2011, pp. 899–909.

[57] C. Germain, Conception d'un robot parallèle à deux degrés de liberté pour des opérations de prise et de dépose, L'Université Nantes Angers Le Mans, 2013.

[58] V.E. Gough, S.G. Whitehall, Universal tyre test machine, in: Proc. FISITA 9th Int. Technical Congress, 1962, pp. 117–137.

[59] J.E. Gwinnett, Amusement device, Google Patents, US Patent 1,789,680, 1931.

[60] A. Hfaiedh, A. Chemori, A. Abdelkrim, RISE controller for Class I of underactuated mechanical systems: Design and real-time experiments, in: Proc. 3rd International Conference on Electromechanical Engineering (ICEE), Skikda, Algeria, 2018.

[61] Hugo Hadfield, Lai Wei, Joan Lasenby, The forward and inverse kinematics of a delta robot, in: Computer Graphics International Conference, Springer, 2020, pp. 447–458.

[62] A. Hfaiedh, A. Chemori, A. Abdelkrim, Observer-based RISE control of Class I of underactuated mechanical systems: Theory and real-time experiments, Transactions of the Institute of Measurement and Control (2022).

[63] K. Harib, K. Srinivasan, Kinematic and dynamic analysis of Stewart platform-based machine tool structures, Robotica 21 (5) (2003) 541.

[64] G. Hassan, A. Chemori, L. Chikh, P.-E. Herve, M. El Rafei, L. Chikh, F. Pierrot, RISE feedback control of cable-driven parallel robots: Design and real-time experiments, in: Proc. 21st IFAC World Congress, Berlin, Germany, 2020.

[65] M. Honegger, A. Codourey, E. Burdet, Adaptive control of the hexaglide, a 6 dof parallel manipulator, in: Proceedings of International Conference on Robotics and Automation, vol. 1, IEEE, 2020, pp. 543–548.

[66] N. Hovakimyan, C. Cao, L1 Adaptive Control Theory: Guaranteed Robustness With Fast Adaptation, SIAM, 2010.

[67] T. Hufnagel, A. Muller, A projection method for the elimination of contradicting decentralized control forces in redundantly actuated PKM, IEEE Transactions on Robotics 28 (3) (2012) 723–728.

[68] R. Kelly, V. Santibáñez Davila, J.A. Loría Perez, Control of Robot Manipulators in Joint Space, Springer Science & Business Media, 2006.

[69] S. Krut, M. Benoit, H. Ota, F. Pierrot, I4: A new parallel mechanism for Scara motions, in: 2003 IEEE International Conference on Robotics and Automation (Cat. No. 03CH37422), vol. 2, IEEE, 2003, pp. 1875–1880.

[70] S. Krut, Contribution à l'étude des robots parallèles légers, 3T-1R et 3T-2R, à forts débattements angulaires, Université de Montpellier 2, 2003.

[71] P.R. Kuma, B. Bandyopadhyay, Stabilization of Stewart platform using higher order sliding mode control, in: 2012 7th International Conference on Electrical and Computer Engineering, IEEE, 2012, pp. 945–948.

[72] A. Lafmejani, M. Masouleh, A. Kalhor, Trajectory tracking control of a pneumatically actuated 6-DOF Gough-Stewart parallel robot using backstepping-sliding mode controller and geometry-based quasi forward kinematic method, Robotics and Computer-Integrated Manufacturing 54 (2018) 96–114.

[73] P. Lambert, J. Herder, A 7-DOF redundantly actuated parallel haptic device combining 6-DOF manipulation and 1-DOF grasping, Mechanism and Machine Theory 134 (2019) 349–364.

[74] Y. Li, Q. Xu, Kinematics and inverse dynamics analysis for a general 3-PRS spatial parallel mechanism, Robotica 23 (2) (2005) 219–229.

[75] Y. Li, Q. Xu, Design and development of a medical parallel robot for cardiopulmonary resuscitation, IEEE/ASME Transactions on Mechatronics 12 (3) (2007) 265–273.

[76] G. Liu, Z. Qu, X. Liu, J. Han, Singularity analysis and detection of 6-UCU parallel manipulator, Robotics and Computer-Integrated Manufacturing 30 (2) (2014) 172–179.

[77] J. Li, C. Qi, Y. Li, Z. Wu, Prediction and compensation of contour error of CNC systems based on LSTM neural-network, IEEE/ASME Transactions on Mechatronics 27 (1) (2022) 572–581.

[78] Y. Lou, Z. Li, Y. Zhong, J. Li, Z. Li, Dynamics and contouring control of a 3-DoF parallel kinematics machine, Mechatronics 21 (1) (2011) 215–226.

[79] Kevin M. Lynch, Frank C. Park, Modern Robotics, Cambridge University Press, 2017.

[80] J.P. Merlet, Parallel Robots, Springer Science & Business Media, 2006.

[81] A. Mueller, Robust modeling and control issues of parallel manipulators with actuation redundancy, in: Recent Advances in Robust Control: Theory and Applications in Robotics and Electromechanics, BoD – Books on Demand, 2011, p. 207.

[82] A. Müller, T. Hufnagel, Model-based control of redundantly actuated parallel manipulators in redundant coordinates, Robotics and Autonomous Systems 60 (4) (2012) 563–571.

[83] V. Nabat, M. de la O Rodriguez, O. Company, S. Krut, F. Pierrot, Par4: Very high speed parallel robot for pick-and-place, in: 2005 IEEE/RSJ International Conference on Intelligent Robots and Systems, IEEE, 2005, pp. 553–558.

[84] V. Nabat, Robots parallèles à nacelle articulée, du concept à la solution industrielle pour le pick-andplace, Université de Montpellier II, 2007.

[85] G.S. Natal, A. Chemori, F. Pierrot, O. Company, Nonlinear dual mode adaptive control of PAR2: A 2-dof planar parallel manipulator, with real-time experiments, in: 2009 IEEE/RSJ International Conference on Intelligent Robots and Systems, IEEE, 2009, pp. 2114–2119.

[86] G.S. Natal, A. Chemori, F. Pierrot, Dual-space control of extremely fast parallel manipulators: Payload changes and the 100G experiment, IEEE Transactions on Control Systems Technology 4 (23) (2015) 1520–1535.

[87] G.S. Natal, A. Chemori, F. Pierrot, Nonlinear control of parallel manipulators for very high accelerations without velocity measurement: Stability analysis and experiments on Par2 parallel manipulator, Robotica 34 (1) (2016) 43–70.

[88] P.R. Ouyang, W.J. Zhang, F.X. Wu, Nonlinear PD control for trajectory tracking with consideration of the design for control methodology, in: Proceedings 2002 IEEE International Conference on Robotics and Automation (Cat. No. 02CH37292), vol. 4, 2002, pp. 4126–4131.

[89] A.S. Ovchinnikov, O.V. Bocharnikova, N.S. Vorobyeva, A.V. Dyashkin, V.S. Bocharnikov, S.D. Fomin, Kinematic study of a robot-weeder with a sprayer function and fertigation, in: IOP Conference Series: Earth and Environmental Science, vol. 422, IOP Publishing, 2020.

[90] B. Paden, R. Panja, Globally asymptotically stable 'PD+' controller for robot manipulators, International Journal of Control 47 (6) (1988) 1697–1712.

[91] F.C. Park, J.W. Kim, Singularity analysis of closed kinematic chains, Journal of Mechanical Design 121 (1) (1999) 32–38.

[92] S.B. Park, H.S. Kim, C. Song, K. Kim, Dynamics modeling of a delta-type parallel robot, in: IEEE ISR 2013, IEEE, 2013, pp. 1–5.

[93] P.M. Patre, W. MacKunis, C. Makkar, W.E. Dixon, Asymptotic tracking for systems with structured and unstructured uncertainties, in: Proceedings of the 45th IEEE Conference on Decision and Control, IEEE, 2006, pp. 441–446.

[94] F. Pierrot, P. Dauchez, A. Fournier, Fast parallel robots, Journal of Robotic Systems 8 (6) (1991) 829–840.

[95] François Pierrot, Robots pleinement parallèles légers: Conception, modélisation et commande, Montpellier 2, 1991.

[96] F. Pierrot, T. Shibukawa, From hexa to hexaM, in: Parallel Kinematic Machines, Springer, 1999, pp. 357–364.

[97] F. Pierrot, C. Baradat, V. Nabat, O. Company, S. Krut, M. Gouttefarde, Above 40g acceleration for pick-and-place with a new 2-dof PKM, in: 2009 IEEE International Conference on Robotics and Automation, IEEE, 2009, pp. 1794–1800.

[98] M. Pham, H. Champliaud, Z. Liu, I. Bonev, Parameterized finite element modeling and experimental modal testing for vibration analysis of an industrial hexapod for machining, Mechanism and Machine Theory 167 (2022) 104502.

[99] F. Plestan, Y. Shtessel, V. Bregeault, A. Poznyak, New methodologies for adaptive sliding mode control, International Journal of Control 83 (9) (2010) 1907–1919.

[100] W.L.V. Pollard, Spray painting machine: USA, Google Patents, US2213108, 1940.

[101] L. Ren, J.K. Mills, D. Sun, Experimental comparison of control approaches on trajectory tracking control of a 3-DOF parallel robot, IEEE Transactions on Control Systems Technology 15 (5) (2007) 982–988.

[102] J.M. Sabater, R. Aracil, R.J. Saltaren, L. Payá, A novel parallel haptic interface for telerobotic systems, in: Advances in Telerobotics, Springer, 2007, pp. 45–59.

[103] N. Sadegh, R. Horowitz, Stability and robustness analysis of a class of adaptive controllers for robotic manipulators, The International Journal of Robotics Research 9 (3) (1990) 74–92.

[104] H. Saied, A. Chemori, M. El Rafei, C. Francis, F. Pierrot, From non-model-based to model-based control of PKMs: A comparative study, in: Mechanism, Machine, Robotics and Mechatronics Sciences, Springer, 2019, pp. 153–169.

[105] H. Saied, A. Chemori, M. Bouri, M. El Rafei, C. Francis, F. Pierrot, A new dynamic-feedback-based RISE control of PKMs: Theory and application, in: Proc. IEEE/RSJ Int. Conf. Intel. Robots and Systems (IROS), Macao, China, 2019.

[106] H. Saied, A. Chemori, M. Bouri, M. El Rafei, C. Francis, F. Pierrot, A new time-varying feedback RISE control for second-order nonlinear MIMO systems: Theory and experiments, International Journal of Control (2019) 1–14.

[107] H. Saied, On Control of Parallel Robots for High Dynamic Performances: From Design to Experiments, Université de Montpellier, 2019.

[108] H. Saied, A. Chemori, M. Bouri, M. El Rafei, C. Francis, F. Pierrot, A new time-varying feedback RISE control for 2nd-order nonlinear MIMO systems: Theory and experiments, International Journal of Control 94 (8) (2021) 2304–2317.

[109] H. Saied, A. Chemori, M. El Rafei, C. Francis, A novel model-based robust super-twisting sliding mode control of PKMs: Design and real-time experiments, in: Proc. IEEE/RSJ International Conference on Intelligent Robots and Systems, Prague, Czech Republic, 2021.

[110] V. Santibañez, R. Kelly, PD control with feedforward compensation for robot manipulators: Analysis and experimentation, Robotica 19 (1) (2001) 11.

[111] Homayoun Seraji, A new class of nonlinear PID controllers with robotic applications, Journal of Robotic Systems 15 (1998) 161–181.

[112] W.W. Shang, S. Cong, Y. Ge, Adaptive computed torque control for a parallel manipulator with redundant actuation, Robotica 30 (3) (2012) 457.

[113] S. Shayya, Towards Rapid and Precise Parallel Kinematic Machines, University of Montpellier, 2015.

[114] K.I.K. Sherwani, N. Kumar, A. Chemori, M. Khan, S. Mohammed, RISE-based adaptive control for EICoSI exoskeleton to assist knee joint mobility, Robotics and Autonomous Systems (2020).

[115] D. Schindele, H. Aschemann, Trajectory tracking of a pneumatically driven parallel robot using higher-order SMC, in: 2010 15th International Conference on Methods and Models in Automation and Robotics, IEEE, 2010, pp. 387–392.

[116] B. Siciliano, L. Sciavicco, L. Villani, G. Oriolo, Robotics: Modelling, Planning and Control, Springer Science & Business Media, 2010.

[117] N. Simaan, Analysis and Synthesis of Parallel Robots for Medical Applications, Technion-Israel Institute of Technology, Faculty of Mechanical Engineering, 1999.

[118] S. Staicu, Dynamics of Parallel Robots, Springer, 2019.

[119] J. Stewart, A platform with six degrees of freedom, Proceedings - Institution of Mechanical Engineers 180 (1) (1965) 371–386.

[120] Y.X. Su, B.Y. Duan, C.H. Zheng, Nonlinear PID control of a six-DOF parallel manipulator, IEE Proceedings. Control Theory and Applications 151 (1) (2004) 95–102.

[121] H.D. Taghirad, Parallel Robots: Mechanics and Control, CRC Press, 2013.

[122] M. Taktak-Meziou, A. Chemori, J. Ghommam, N. Derbel, A prediction-based optimal gain selection in RISE feedback control for hard disc drives, in: Proc. IEEE Conference on Control Applications (CCA), Antibes, France, 2014.

[123] M. Taktak-Meziou, A. Chemori, J. Ghommam, N. Derbel, RISE feedback with NN feed-forward control of a servo-positioning system for track following in HDD, Transactions on Systems, Signals and Devices (2015).

[124] M. Taktak-Meziou, A. Chemori, J. Ghommam, N. Derbel, Mechatronics of hard disk drives: RISE feedback track following control of a R/W head, in: Mechatronics: Principles, Technologies and Applications, Nova Science Publishers, 2015.

[125] W. Tian, F. Yin, H. Liu, J. Li, Q. Li, T. Huang, D. Chetwynd, Kinematic calibration of a 3-DOF spindle head using a double ball bar, Mechanism and Machine Theory 102 (2016) 167–178.

[126] L.W. Tsai, Robot Analysis: The Mechanics of Serial and Parallel Manipulators, John Wiley & Sons, 1999.

[127] R. Ur-Rehman, S. Caro, D. Chablat, P. Wenger, Multi-objective path placement optimization of parallel kinematics machines based on energy consumption, shaking forces and maximum actuator torques: Application to the Orthoglide, Mechanism and Machine Theory 45 (8) (2010) 1125–1141.

[128] V. Utkin, H. Lee, Chattering problem in sliding mode control systems, in: International Workshop on Variable Structure Systems, 2006, VSS'06, IEEE, 2006, pp. 346–350.

[129] S.T. Venkataraman, S. Gulati, Control of nonlinear systems using terminal sliding modes, in: 1992 American Control Conference, 1992, pp. 891–893.

[130] M. Wapler, V. Urban, T. Weisener, J. Stallkamp, M. Dürr, A. Hiller, A Stewart platform for precision surgery, Transactions of the Institute of Measurement and Control 25 (4) (2003) 329–334.

[131] Wikipedia contributors, Full flight simulator, Wikipedia, The Free Encyclopedia, https://en.wikipedia.org/w/index.php?title=Full_flight_simulator&oldid=1041982466, 2021.

[132] Robert L. Williams II, The Delta Parallel Robot: Kinematics Solutions, Mechanical Engineering, Ohio University, October 2016.

[133] B. Xian, D.M. Dawson, M.S. de Queiroz, J. Chen, A continuous asymptotic tracking control strategy for uncertain nonlinear systems, IEEE Transactions on Automatic Control 49 (7) (2004) 1206–1211.

[134] B. Xian, M.S. Queiroz, D.M. Dawson, A continuous control mechanism for uncertain nonlinear systems, in: Optimal Control, Stabilization and Nonsmooth Analysis, Springer, 2004.

[135] H. Yang, L. Chen, Z. Ma, M. Chen, Y. Zhong, F. Deng, M. Li, Computer vision-based high-quality tea automatic plucking robot using Delta parallel manipulator, Computers and Electronics in Agriculture 181 (2021) 105946.

[136] D. Zhang, Parallel Robotic Machine Tools, Springer Science & Business Media, 2009.

[137] Y.J. Zhao, Z.Y. Yang, T. Huang, Inverse dynamics of Delta robot based on the principle of virtual work, Transactions of Tianjin University 11 (4) (2005) 268–273.

[138] Y. Zhao, F. Gao, Inverse dynamics of the 6-dof out-parallel manipulator by means of the principle of virtual work, Robotica 27 (2) (2009) 259–268.

[139] H. Zhuang, Y. Wang, A coordinate measuring machine with parallel mechanisms, in: Proceedings of International Conference on Robotics and Automation, vol. 4, IEEE, 1997, pp. 3256–3261.

[140] B. Zi, N. Wang, S. Qian, K. Bao, Design, stiffness analysis and experimental study of a cable-driven parallel 3D printer, Mechanism and Machine Theory 132 (2019) 207–222.

[141] J.G. Ziegler, N.B. Nichols, Optimum settings for automatic controllers, Transactions of the American Society of Mechanical Engineers 64 (11) (1942).

Index

Printed in the United States
by Baker & Taylor Publisher Services